Prevention of Premature Staining of New Buildings

Prevention of Premature Staining of New Buildings

Phil Parnham

Principal Lecturer in Building Surveying, Sheffield Hallam University, UK

E & FN SPON

An Imprint of Chapman & Hall

London · Weinheim · New York · Tokyo · Melbourne · Madras

Published by E & FN Spon, an imprint of Chapman & Hall, 2–6 Boundary Row, London SE1 8HN, UK

Chapman & Hall, 2–6 Boundary Row, London SE1 8HN, UK

Chapman & Hall GmbH, Pappelallee 3, 69469 Weinheim, Germany

Chapman & Hall USA, 115 Fifth Avenue, New York NY 10003, USA

Chapman & Hall Japan, ITP-Japan, Kyowa Building, 3F, 2-2-1 Hirakawacho, Chiyoda-ku, Tokyo 102, Japan

Chapman & Hall Australia, 102 Dodds Street, South Melbourne, Victoria 3205, Australia

Chapman & Hall India, R. Seshadri, 32 Second Main Road, CIT East, Madras 600 035, India

First edition 1997

Typeset in 10/12pt Times by Cambrian Typesetters, Frimley, Surrey

Printed in Great Britain by St Edmundsbury Press, Bury St Edmunds, Suffolk

ISBN 0 419 17130 4

A catalogue record for this book is available from the British Library

Contents

For Sue, Laura and Jennie

Preface

A local authority had completed a mixed low-rise housing scheme in an inner city area. It was an ambitious project and represented the last new build scheme carried out before central government cutbacks of the 1980s brought an end to council house-building programmes. It was a small estate of 132 traditionally built dwellings completed in 1986 to replace a 'sink' estate of walk-up flats. The development was reviewed in 1988 by the *Architects Journal* (Hannay, 1988).

Despite the pride and excitement surrounding the scheme's completion, the visual image of the buildings did not stand the test of time. Within a year of completion, ugly algae staining had affected almost every constructional detail on many elevations. The balconies, external staircases, window cills and parapet walls were most badly affected (see Figure 0.1). The visual effect was so overwhelming that the disfigurement became the most dominant feature of the development to many occupants and observers.

Discussions with the design team and closer inspection revealed that one of the design approaches was to create buildings with 'clean lines' omitting all overhangs, drips, throatings, etc. This approach was encapsulated by the detailing of the boundary walls to some parts of the site. Creasing tiles had been included beneath the brick on edge copings as a reference to vernacular detailing that was designed to minimize the flow of water down the face of the wall. However, to avoid the overhanging tiles spoiling the cohesion of the design philosophy, the contractor was instructed to cut back the tiles flush to the brickwork. This resulted in the uncontrollable flow of rainwater over the face of the wall and, where this had become concentrated, dirt staining and algae growth quickly became established on the brickwork (see Figures 0.2, 0.3 and 0.4).

Reaction to this problem was surprisingly mixed. The tenants and the local housing managers were upset that such a new development had been spoiled so early in its life. Yet the design professionals did not view it as being so significant, especially when placed in context with other more successful features of the scheme. Hannay (1988, p. 48) paid little attention to the problem.

Figure 0.1 Elevation of housing development affected by premature staining.

Figure 0.2 The boundary wall was particularly badly affected.

Figure 0.3 Detail of Figure 0.2.

Following a thorough assessment of all aspects of the scheme, Hannay noted: 'The odd weathering detail lets the otherwise pristine elevations down.'

This case raises two important issues:

- The aesthetic performance of a building façade is just one component part of the much broader *design problem* that a designer has to resolve when designing any building.
- The criteria used by the public and other non-design professionals to judge the success of a finished building can be very different from those used by the designers.

This last point is a reasonable one. The real test of a housing scheme should be whether it offers such features as enough space for the occupants, whether it creates the right sort of internal environment and results in a harmonious external environment that users find convenient, pleasant and safe. But to many people, especially the infrequent user or casual observer, the overall design success of this scheme failed to outweigh the lasting impression of the ugly stains running down the front of it.

EXTENT OF THE PROBLEM

Reviewing what published material exists on this topic and observing several hundred new buildings, it is clear that this is not an isolated case but is representative of a quite widespread problem. Therefore it cannot be the fault of a few poorly trained or thoughtless individual designers but evidence of a fundamental lack of awareness across the design profession as a whole. Atkinson (1983) considered this issue and proposed that, although

Figure 0.4 Detail of brick coping showing flush tile creasing.

an understanding of the nature of building materials is the keystone of architectural design, much of the existing analysis has been focused on how buildings degrade and decay while 'changes in appearance over time, and the acceptability of a change to the building owner, passer by and to architects and critics, have seldom been matters for research.'

NO EASY ANSWERS

It would be easy to be wholly critical of the designers of new buildings, the opportunities being so numerous. But what this study reveals is that striking the right balance between all the competing components in a design problem is not easy, especially when the building has to be completed on time, within cost limits and to a standard that will please the client. For those who take pleasure in criticizing defects and failures in new buildings, remember the old adage:

HINDSIGHT IS AN EXACT SCIENCE

OBJECTIVES OF THIS BOOK

When the problem of premature staining of new buildings was first considered, a number of key questions were identified:

- Why has staining become so prominent?
- What causes premature staining?
- What parts of new buildings are most likely to be affected?
- How can it be avoided?

The book has been organized to address these questions in a very practical and accessible way, allowing it to be used as a quick drawing-board reference book for designers in a rush while catering for those who want to take a broader view. So this book aims to achieve a number of things:

- to amalgamate, extend and update previous work done in this area;
- to make designers aware of the factors that lead to the premature staining of their buildings;
- through in-depth study of a number of examples, to propose an approach that will help designers to think how their buildings will perform over time and so minimize rapid and adverse changes in appearance;
- to propose a number of standard details that might help in achieving these aims.

REFERENCE

Hannay, P. (1988) View from the top. *Architects Journal*, **188**(42), 42–53.

FURTHER READING

Atkinson, G. (1983) Appearance matters. *Building*, **294**(19), 64–65.

Acknowledgements

The author would like to acknowledge the contributions and assistance given during the writing of this book. Although they are too numerous to mention, a few are listed below:

- the Educational Trust of the Royal Institution of Chartered Surveyors for the grant that initially supported this work;
- colleagues and fellow building surveyors in the School of Construction at Sheffield Hallam University for all their advice, guidance and support;
- Andy Catling for his advice on trees;
- and all my former colleagues, some of whom provided the raw material for this book!

Chapter 1
Defining premature staining

1.1 DEFINITION OF PREMATURE STAINING

Before the mechanics of staining are considered in greater depth, it will be worth while reviewing how the title of this book was arrived at. When considering the changing appearance of buildings over time the term 'weathering' initially seemed the most appropriate. This is defined in *Chambers Science and Technology Dictionary* as:

> The process of disintegration and decomposition effected in minerals and rocks as a consequence of exposure to the atmosphere and to the action of frost, rain and insolation. These effects are partly mechanical, chemical and organic and their continuation depends upon the removal, by transportation, of the products of weathering. (*Walker, 1988*)

This is a geological definition that describes the processes acting on the earth's crust; in a building context the term is usually associated with the changes in building materials due to the action of the rain, wind and sun. Addleson and Rice (1994) used the term to include not only superficial changes in appearance but also the decay and disintegration of the materials used externally. As this would affect all buildings regardless of their design or construction and include the physical breakdown of materials, its scope becomes too broad for this study.

Another possible term was 'soiling' of building façades but as this has a dictionary definition of 'To make dirty especially on the surface', it suggests a temporary change in appearance that can easily be removed. The subject matter of this book is really something more permanent, which makes the term 'staining' more applicable as this is 'a mark or discoloration that is not easily removed'. Applying other criteria, the central concept of this study can be defined:

- *Prevention* – to keep from happening.
- *Premature* – something occurring or existing before its normal or expected time.
- *Staining* – a mark or discoloration that is not easily removed.
- *New* – recently made, brought into being or acquired.
- *Buildings* – any structure that has a roof and walls.

Using these criteria, 'prevention of premature staining of new buildings' can be further defined as:

> The avoidance of discoloration or disfigurement of the façade of a new building before the time that such a change in appearance can be expected.

Premature staining is not the patina of age, that gradual progression through youth, maturity and old age that most buildings will inevitably experience. Instead it is a dramatic soiling of the face of the building within the first few years of its life.

The problem will normally be isolated and localized, associated with specific design features or changes in materials producing a visual effect that the designer did not anticipate or intend. As Partridge (1971) stated when discussing how an original design intention cannot be divorced from weathering detail considerations:

> There is always the chap (*sic*) who wants an unusual colour for his building – I have met many of them. There could conceivably be the chap who intended his building to be white and covered with black streaks – I have not met him yet!

Although the agents that cause staining are the same as those responsible for the general weathering of external surfaces, they act in a concentrated fashion to accelerate a change in appearance.

Premature staining can be considered as failure in the aesthetic performance of the façade and the examples in this book cannot be necessarily identified as *building defects*. This term suggests a physical fault in the design or construction of a part of a building, a feature or omission that leads to structural failure, cracking, water penetration or other failure in performance. Although premature staining can be the first stage in the physical breakdown of the façade, its initial impact is visual. Because of this, many within the design and construction industry do not see it as a *defect* and so attach little importance to it. If more people did, then perhaps this book would not be necessary.

In terms of the time scale of the staining, most of the examples in this book are from buildings that are less than five years old. Older buildings have been included where they provide good illustrative examples.

1.2 PREMATURE STAINING VS AGEING

In most climates atmospheric pollutants and biological growths will combine with the actions of the wind and rain to change the appearance of building materials over time. Although this change is unavoidable the nature and extent of it will vary according to a wide range of factors that cannot be accurately predicted. Façades of buildings will go through their own stages of youthfulness, middle and old age with each exhibiting their own characteristics. According to Hawes (1986), many of the buildings designed over the last 50 to 60 years have failed to acknowledge this inevitability, being 'designed only with youth in mind and too many have proved quite incapable of gracefully accepting the imprint of the passing years.'

Passage of buildings into the third age, where many become 'historic', the accumulation of dirt may in fact add to the visual authority of the building through this noble patina of age. This concept has been illustrated by Carrié and Morel (1975) in their influential book *Salissures de façades*. A photograph of the porch at the Church of St Margaret at Westminster shows years of accumulation of London grime yet this is not resented by the thousands of people who visit the building every year (Figure 1.1). But what happens when it is cleaned (Figures 1.2 and 1.3)? Does it lose any of its nobility? Studies carried out in Scotland have suggested that the public would feel important buildings would lose their character if dirt was cleaned from the façade. In this case Edinburgh Castle was the building under consideration (Andrews, 1994, p. 38). This highlights the complex relationship between premature staining and the normal ageing process of buildings, a subject that will be considered in more depth in Chapter 4.

Figure 1.1 The church of St Margarets, Westminster. Soiled porch before cleaning (Hawes, 1986).

1.3 THE EFFECTS OF PREMATURE STAINING

The disfigurement of a building's façade can have profound implications for all those involved in its design, construction, ownership and use, as given below.

1.3.1 Economic implications

The *value* of the building may well be affected if the staining is significant, both directly by a reduction in the valuation of the property and indirectly by being less attractive to purchasers or lessees. The scale of this effect is difficult to quantify as little research work has been carried out in this area. R.A. Laing, a research student at Robert Gordon University, investigated the likely impact on value brought about by cleaning masonry buildings in Scotland (private correspondence). A survey of general practice surveyors across eight Scottish cities suggested that the alteration in valuations following cleaning operations would be marginal and at the most optimistic show a 1–2% increase in value. However, a large number of respondents considered that the marketability of a property would be enhanced if the building was cleaner and so more attractive. In a time of recession such advantages may be significant as lower marketing costs, lower advertising expenses and a quicker return on capital may all represent a substantial financial benefit. Although this study focused on older masonry buildings, the principle may hold true for new buildings that have been prematurely stained. The commercial development illustrated in Figures 4.63–4.72 shows a group of new buildings that were badly disfigured within months of being completed and certainly before they were let or sold. The buildings stood empty for several months and, given that there was a glut of office accommodation on the market at the time, it is conceivable that ugly algae- and dirt-stained façades might be the deciding factor that causes a potential leasee or purchaser to look elsewhere.

1.3.2 Prestige

A stained façade dramatically reduces the visual impact of what would otherwise be a prestigious building. Where corporate organizations have commissioned bold and expensive buildings to reflect their ethos, dramatic disfigurements will affect the image and reputations of the designers, constructors but mainly the occupiers. Some of the buildings shown in this book house organizations that have a high public profile. Some were contacted for their views on the problems and were at best 'succinct' in their reaction, at worse downright aggressive.

Figures 1.2 and 1.3 The porch of St Margarets after cleaning.

1.3.3 Impact on the public and users

In recent times, the important link between the effects of a building or a structure on the behaviour of its users or passers-by has long been recognized. In relation to stained façades, Dr C. Steffen, a psychologist with the Department of Architecture at the Delft University of Technology, conducted one of the few research projects into people's appreciations of the façades of buildings (Verhoff, 1988, p. 61). Although the study was restricted to one building only, the results revealed that a person's experience of clean versus dirty façades was based on six variables:

- how businesslike they were;
- how orderly;
- how interesting;
- how complex;
- how stimulating;
- and how safe they were felt to be.

Verhoff commented that repeated investigations of different buildings should be carried out to validate this study and that they should concentrate on the following topics:

- *Observation* – how soiled must a façade be before it is noticed by an arbitrary observer?
- *Cognition* – how much can a façade be soiled before there is a need for information on the origins of the soiling and its physical effects on the façade?
- *Affective value experience* – what emotions arise from a soiled façade? When is a façade so soiled that it is considered ugly, repulsive and depressive? In contrast, is there a point where the soiled façade is so dirty that the characteristic of the total form is considered beautiful, fascinating and fine?
- *Motivation* – what is the psychological value of a clean façade for the user and the owner of the building?

- *Behaviour* – what type of behaviour can be caused by a soiled façade?

Verhoff considered that the results of such a study would be of immense value to designers who could begin to consider their designs over time.

This area of research has been carried forward by Andrews (1994), working as part of the Masonry Conservation Research Group at the Robert Gordon University, who sought to investigate people's reaction to the cleaning of older masonry buildings in Scotland. Photographs were taken of a number of buildings before and after cleaning and subjects asked what they felt about the result. They were asked to express their opinion in relation to a series of predetermined semantic differentials or opposing terms. These are outlined below:

1. Well looked after	Shabby
2. Impressive	Unimpressive
3. Delicate	Weighty
4. Distinctive	Ordinary
5. Inviting	Repelling
6. Orderly	Irregular
7. Cheerful	Gloomy
8. Warm	Cold
9. Attractive	Unattractive
10. Delightful	Dreadful
11. Has character	Has no character
12. Soft	Hard
13. Clean	Dirty
14. Tidy	Untidy
15. Friendly	Unfriendly
16. Light	Dark
17. Pleasing colour	Displeasing
18. Elegant	Clumsy
19. Uplifting	Depressing
20. Dignified	Undignified
21. High status	Low status
22. Unique	Common

The graph in Figure 1.4 shows the change in people's evaluation of the buildings after cleaning, using these headings. The reference numbers of these differentials are represented along the bottom axis of this graph. The greater the percentage change in the category the greater the impact that cleaning had. The results can be divided into groupings in descending order of importance:

- *Group One* – clean–dirty; light–dark;
- *Group Two* – pleasing colour; cheerful; tidy; well looked after;
- *Group Three* – attractive; warm, soft; friendly; elegant; uplifting;
- *Group Four* – orderly; character; unique; impressive; weighty.

The conclusions drawn from this study give a useful insight into owners', users' and the general public's view of buildings that may (or may not) prematurely stain. These include:

- Soiling is one of the most important ways buildings are conceptualized by people.

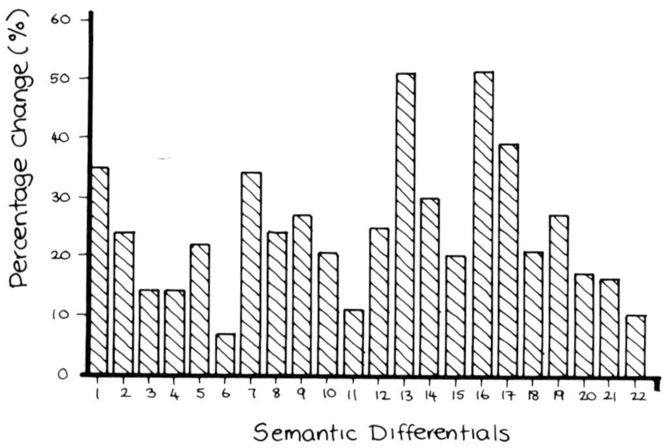

Figure 1.4 Changes in people's evaluation of buildings after cleaning.

- Soiling may not be considered so important if the level and distribution of the disfigurement is consistent with the underlying material and architectural detail (i.e. is not unsightly as in the case of Figures 0.1–0.4 in the Preface).
- Cleaner buildings are perceived more positively than equivalent soiled ones over a broad range of characteristics.

It must be acknowledged that this study concentrated exclusively on stone-built older buildings and these conclusions may not be directly applicable to modern buildings with different architectural styles. Further aspects of this important work are discussed in Chapter 5.

1.3.4 Effect on the fabric of the building

Although premature staining has been defined as not being a building defect, in reality the mechanisms involved in staining are the same as those involved in the degradation of the façade material. Water is always associated with the staining process and enhanced material breakdown through frost and chemical action. Therefore these processes are really on a continuum with aesthetic alteration at one end and material breakdown at the other.

REFERENCES

Addleson, L. and Rice, C. (1994) *Performance of Materials of Buildings*, Butterworth–Heinemann, p. 144.

Andrews, A. (1994) *Stone Cleaning. A Guide for Practitioners*, Historic Scotland, Edinburgh, p. 38.

Carrié, C. and Morel, D. (1975) *Salissures de façades*, Editions Eyrolles, Paris, p. 140.

Hawes, F. (1986) *The Weathering of Concrete Buildings*, Cement and Concrete Association, Slough, p. 2.

Partridge, J.A. (1971) Architectural design and detailing. The weathering of concrete. *Concrete Society Symposium*, Cement and Concrete Association.

Verhoff, L.G.W. (ed.) (1988) *Soiling and Cleaning of Building Façades*. Report of the Technical Committee 62 SCF, RILEM, Chapman and Hall, London, p. 61.

Walker, P.M.B. (ed.) (1988) *Chambers Science and Technology Dictionary*, W. & R. Chambers Ltd.

Chapter 2
Historical context

2.1 INTRODUCTION

Buildings that have stained prematurely can stir the emotions, especially of those who have responsibility for their ongoing care and maintenance. Ingwell (1985) made a plea for the reintroduction of projections on modern buildings so that 'we don't have to endure the ugly and deleterious effect of pattern staining, mould, algae and lichen growth etc., on some of our new buildings'. He claimed that our Georgian and Victorian forebears knew of the need of projections and 'tendency to omit projection courses such as cornices, string courses and sills – with their drips and throatings – has resulted in many recent buildings becoming horribly disfigured within a very short time after erection.' Declining a return to gargoyles and corbels, he appealed for more thought to be given to detailing in this area.

So have we lost the detailing skills that enabled buildings to resist premature staining? Is it the fault of modern architectural fashions? Did the Georgians and Victorians really have an insight into the staining process or is that a rose-tinted nostalgic look back to simpler times? Part of the answer to these questions can be found in a brief historical review of architectural design this century.

2.2 BATTLE OF THE STYLES

History reveals that this is not a new issue; similar debates were raging nearly 100 years ago. Towards the end of the 19th century, new building designs were generally characterized by wide cornices, belt courses and heavy detailing. Known as the 'battle of the styles', devotees of Classicism, Romanesque and Gothic promoted their preferences and developed their own personal designs based on historical examples. Consequently, buildings of the Industrial Revolution became clothed in one or other of these styles. Architectural reference books of the era were clear on this issue. In a reference book produced by the professional staff of the Bennett College in Sheffield for the use of their students, it was stated that modern architecture

> is mostly based on one or other of the two traditional schools of design known as Gothic and Classic. . . . To ensure becoming an expert draughtsman, it is necessary to have a practical and ready insight of the main details and principles of at least both of these schools of design.

Departure from these strict rules was allowed but only as long as the designer adhered to 'the main proportions, in the strict sense and treat minor details in a broader manner . . . in order to meet modern requirements in the planning or scheming of buildings especially of the commercial class.'

2.3 THE DEFEAT OF ORNAMENT

For some in the profession, these strict rules which led to the indiscriminate application of ornament to buildings was without justification and eventually there was a clear reaction against it. Collins (1965) identified the year 1908 as a turning point in architectural design when Adolf Loos published an article entitled 'Ornament and Crime' in which he put the case against artificial application of ornament. Collins attributed Loos with the singular distinction of being responsible both for the plainness of architectural surfaces after this date as well as the abolition of architectural ornament itself. The intricate details that adorned so many of the late 19th-century buildings literally ceased to exist after Loos' article was published. Construction technology books of the era reflected this criticism. In an unattributed book published at the turn of the century, the topic of 'Expedients for throwing rain off walls' was described. Various moulded string courses with weathered surfaces and undercut drips were promoted as features that would protect the walls below from dampness. At the end of the section the commentary somewhat defensively adds: 'moulded string courses, therefore, are not mere ornamental superfluities; they are ornamental but they are also useful' – possibly the words of an ornamentalist under attack.

Not all architects of this era were set against ornament. Augustus Perret, a pioneer in the use of concrete in buildings stood out against this new influence. In 1934, he replied to a questionnaire on the use of ornament in the periodical *Beaux-Arts*. Entitled 'For and against ornament', his article stated:

Figures 2.1 and 2.2 Typical examples of Perret's buildings in Le Havre. Note the strong overhanging horizontal/string courses.

> Contemporary façades are too naked, and yet those which give this effect are generally clothed in plaster rendering or facings, and are thus more denuded than nude. . . . Let us give back to our buildings the organs necessary for their defence against the weather: cornices, string courses, architraves and mouldings, which allow a façade to remain what the artist intended it to be, in spite of the rain. (*Collins, 1959, p. 256*)

Le Corbusier, a designer responsible for many 'nude' buildings, put Perret's attitude down to either his indifference to sociology or to an inability to visualize architecture except in terms of a traditional façade (Collins, 1959, p. 256).

Perret's buildings show this influence and many of the post-war buildings in Le Havre in Northern France were masterminded by him. A number of these were ornamented with horizontal string courses to most storeys providing features that have arguably helped these concrete buildings avoid some of the disfiguring visual effects of some of their contemporaries (Figures 2.1 and 2.2).

2.4 THE MYTH OF THE TRADITIONAL DETAIL

The move away from building details of the Victorian era has traditionally been seen as being responsible for many of the problems of staining. Closer examination of these 'styles' and the approach to design reveals that the guiding principles were historical rather than physical. The Bennetts School guide accepts the need to account for climatic influences and 'a certain amount of forethought must be exercised by the designer of buildings in the provision of boldly projecting porches, strong roofs of moderate pitch and strong walls . . . all designed to protect the buildings from the weather' but still recommended details that held true to historical influences rather than their ability to shed water (Figure 2.3).

If buildings of this period have not been disfigured by staining then this is possibly due to the façades' ability to

'carry' dirt accumulation without creating an adverse visual impression rather than the inherent ability of the ornamental form to prevent staining. This relates to how people perceive buildings in their environment and is discussed in more detail in Chapter 5.

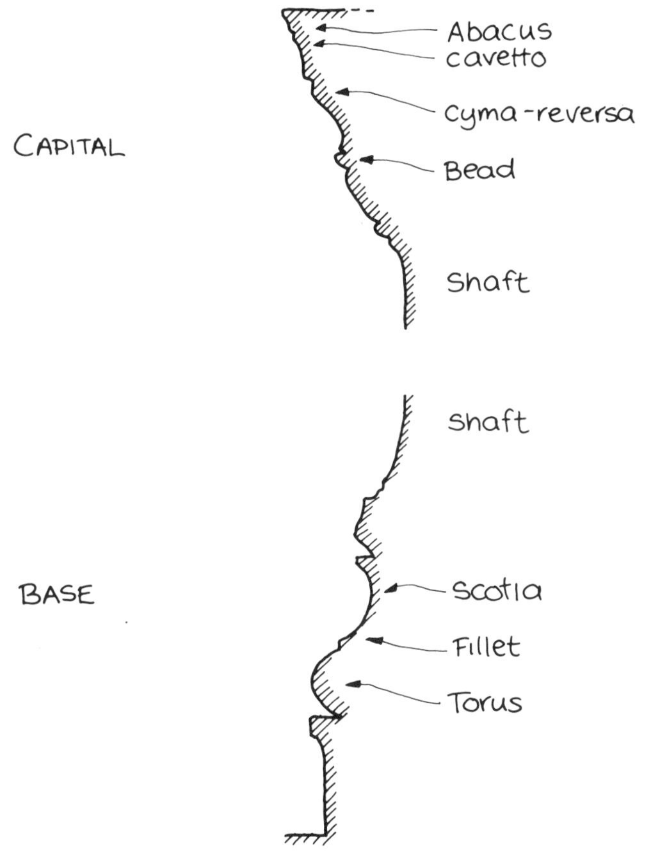

Figure 2.3 Typical ornamental details. This illustration shows details through a column following a classic style. Although heavily ornamented, few of these details would throw flowing water clear of the building.

2.5 THE CONTEMPORARY DEBATE

The Building Research Establishment also noted this move away from traditional practices in architecture. The search for a new architecture in the 1930s or 'modernism' was blamed for the avoidance of pitched roofs and the use of cement-rendered walls. The increased use of the flat roof in particular led to further exposure of the walls:

> . . . some of these early experiments in modern architecture are now familiar by their dated and frequently shabby appearance bought about by the all too hasty abandonment of traditional practices based on long experience. (*BRE, 1964*)

It was also noted that the industry faced such an immense reconstruction programme in the early 1960s that both new and old materials were being used in new bewildering ways: 'Unless these choices are carefully thought out, the repercussions on design detailing would be serious.'

More recently Tony Fitzpatrick, a director of Ove Arup and Partners, commented on one of the materials that has the worst reputation in relation to staining – concrete. At a conference called 'Creativity in Concrete', he stated:

> To be honest, anyone can do wonders with concrete in Barcelona. You can draw something and it stays the same. But if you try to do the same thing here, it will not look the same thing in five years time. . . . What is needed is a process of education in the detailing of concrete. (*Ridout, 1989*)

At the same conference, David Rock of Rock Townsend agreed stating:

> In terms of straightforward architectural knowledge, detailing concrete to resist weathering is the same as for Portland Stone or brickwork. It is a case of remembering classical detailing, and I don't mean Classical Architecture, and including details such as cornices to throw the rain off. These can be incorporated in modern design detailing.

He added that the knowledge should be there but the present architectural education system has lost sight of this.

But it is not just as simple as that. The abandonment of the Classical styles and failure of the educational system cannot be blamed solely for these changes. Atkinson (1977) identified two factors that make the job of the modern architect far more difficult than his or her predecessors:

- the scale of many modern buildings, and
- the size of the façade units used.

The New York World Trade Centre is given as an example. The façade of this building is over 28 000 m^2 with each cladding panel with an area of over 10 m^2. Contrast this with St Peter's in Rome. This building has a façade area of just 5000 m^2 and a panel size of just 0.75 m^2. With such a difference in scale, it is hardly surprising that one building can 'carry' the visual results of weathering better than the other. Such large components also pose considerable technical problems as well.

2.6 CONCLUSIONS

This short historical sketch has revealed a number of dilemmas in regards to the design of modern buildings. The growth of the modern movement and its influence on the design of buildings this century has fundamentally changed the approach to the detailing of façades. The result is a tendency towards plain elevations where overhanging and projecting features have been minimized which still remains influential even in this post-modernist era.

The ornamented detailing of the Victorian period is still considered by many (usually non-designers) as being the product of an inherent understanding of how a building should be weathered – a skill that many commentators blame the modern movement for extinguishing. But analysis of this stylistic detailing reveals that it was itself as much a product of an architectural fashion as the modern movement was.

New buildings cannot be compared to those built 100 years ago as their form and function are so completely different that traditional rules and approaches are just not applicable. Another factor that affects the way modern buildings are designed are the procurement systems employed by building owners. Design-and-build contracts are generally driven by the priorities of contracting organizations where the design professionals are brought in as specialists to advise on the design-related issues of the project. This could further undermine any input that might have stood a chance of avoiding staining problems.

As the old proverb states, 'history will repeat itself', and one of the most effective strategies of minimizing staining will be to learn from the techniques and methods of the past and update and apply them in a contemporary context.

REFERENCES

Atkinson, G.A. (1977) External vertical surfaces of buildings: aspects of design and appearance. *Evaluation of the Performance of External Vertical Surfaces of Buildings*, RILEM, Espoo, Finland, pp. 19–25.

BRE (1964) *Design and Appearance – 1. Building Research Establishment Digest No. 45*. BRE, Garston.

Collins, P. (1959) *Concrete: The Vision of a New Architecture*, Faber.

Collins, P. (1965) *Changing Ideals in Modern Architecture 1750–1950*, Faber.

Ingwell, A.R. (1985) A stain on our buildings. *Building Technology and Management*, **23**(9), 26.

Ridout, G. (1989) Streaking behind. *Building*, **254**(48), 73.

Chapter 3
How and why buildings stain

3.1 INTRODUCTION

In order first to understand why buildings are affected by premature staining and then to develop design skills that help to minimize this disfigurement, it is necessary to understand the mechanisms that cause it. This chapter does not take an in-depth view of how different building materials react to atmospheric and weathering agents; that has been covered much more competently in other publications. Instead it outlines the framework of the topic with enough guidance on further reading for those who want to study the topic in more depth.

3.2 THE ESSENTIAL MECHANISMS OF STAINING

The staining profile of a building façade will be determined by the interactions of three main elements:

- staining agents – these include biological and non-biological agents;
- the characteristics of the materials of the façade;
- the action of the rain and the wind.

The essential process of staining is initiated by water run-off. Without it, the staining agents (or dirt) would probably affect the face of the building evenly. When water does flow down a vertical surface, dirt particles are picked up and redistributed to new positions, usually unevenly. Addleson and Rice (1994, p. 565) put forward the analogy of a river or stream, where swiftly flowing water will pick up solid particles and deposit them when the rate of flow recedes. This effect is illustrated in Figure 3.1 and the remainder of this chapter will consider the process in more detail.

3.3 SOILING AGENTS

There are two main types of soiling agents that affect building façades: biological and non-biological.

3.3.1 Biological soiling

Given the right conditions, many building materials will be affected by biological colonization and the change in appearance it can cause can often be mistaken for soiling by mineral deposition. Although the organisms cannot be seen with the naked eye, the soiling produced can be quite visible and often striking. The buildings in Figures 3.2 to 3.5 show this effect.

Types of organism. There are three main types of organism that can be found on façades: bacteria, microscopic algae, and fungi. The characteristics of each type are outlined below.

- *Bacteria* – these are organisms that exist as living cells, isolated or in colonies and will be present everywhere but originally come from the soil or water. They are unlikely to be visible as surface growths but are capable of causing discoloration as well as deterioration of stone and corrosion of metals. Sulphur-oxidizing bacteria can be involved in the corrosion of iron and steel as well as attacking the calcareous binders of limestone and sandstone, and nitrifying bacteria will affect mortars, concrete and other cement-based products.
- *Algae* – these free-living plants derive their energy mainly from the sunlight. They are present in the air as

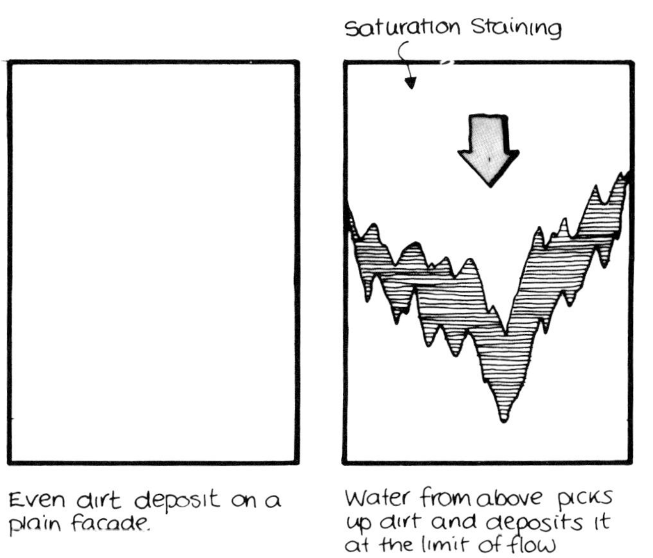

Figure 3.1 Illustration showing partial washing and depositing effects of flowing water over a façade material.

Figure 3.2 General view of a building clad with smooth faced stone panels possibly sandstone.

Figure 3.3 Detail showing extensive algae growth to the surface of the cladding around the base of the building. Here water flow from the top of the parapets and hardstanding has been directed over the face of the panels. The elevation faces north.

spores, originating in the soil and becoming airborne with dust. There are many different types that can tolerate a range of alkaline and acid substrates; they are most commonly green but can also cause red, brown or black stains. They are predominant on surfaces that are frequently wetted and remain damp for some time afterwards. Although they need moisture to thrive, many algae can resist dry conditions for up to a year and still grow once sufficient humidity returns. Algae secrete acids which can dissolve calcium carbonates and can also act on surfaces by the insertion of cells into the pores of the base material. These cells will often exert mechanical pressure through wetting and drying cycles which can lead to cracking of the substrate. Even where algae growth dies back they can serve as a basic nutrition for further growths of bacteria and fungi.

- *Fungi* – these simple plants get their energy from other organic matter. They need moisture but can grow without sunlight and can secrete organic acids which may attack the substrate. Some of these enzymes are very powerful and can break down paint films and thick plastic coatings.

Other biological growths. As the growth of bacteria, algae and fungi becomes established over time, they can modify and lead to the development of other plants. For example, *lichens* can develop from the symbiosis of certain algae and fungi. These are leathery encrustations on the surface that may be embedded in the material itself. They often obscure inscriptions and carvings and will trap water that may cause damage through the freeze–thaw cycle. *Mosses* are green cushions of low spiky tufts that spread outwards and are usually brown when dry. They are most common on surfaces where salts, soil and dirt have accumulated and so are often found in gutters, on roofing materials and very damp brickwork.

More advanced stages of deterioration often provide a footing for the development of ferns and even higher botanical species. Small elder bushes are often seen growing out of mortar joints at high level with roots extending far into the structure. Although these growths are not responsible for premature staining of newer buildings, they represent a second phase of biological colonization that has long-term maintenance implications for the building owner.

Conditions for growth. Generally speaking, conditions for micro-organism development depend on several factors:

- pH value of substrate
- presence of moisture
- temperature
- light
- nutrients

Figure 3.4 Part elevation of a stone-clad building in Glasgow.

Figure 3.5 Detail of Figure 3.4. Water was allowed to flow off the metal roof and onto each glazed area cumulating as it travelled downwards. There was no gutter or drain to minimize the flow which resulted in extensive algae growth increasing in intensity at lower levels.

As there is such a wide range of species in each category, growths can be supported on most façade surfaces. In the UK conditions are especially favourable on the wetter, northern aspects of buildings though lichens can flourish on south-facing elevations.

Planted areas and trees. Particularly healthy biological growths can be found near external planting areas to buildings which offer a rich source of nutrients. Overhanging bushes, shrubs and rich humus soils all provide the nourishment that algae, lichens and mosses need. Shaded by foliage and often well watered, low parapet walls around planted areas are particularly vulnerable.

Overhanging trees can also contribute significantly to staining problems. Useful guidance on this and other aspects involving buildings and trees is contained in a Forestry Commission publication, *Urban Forestry Practice* (Hibberd, 1989). This highlights the varieties of trees that can create problems for adjacent buildings. A few of the main culprits have been described below.

- Beech – a woolly aphid occurs on the underside of the leaf and a combination of sticky honeydew and curled leaves can pose a problem on adjacent surfaces.
- Limes – certain limes are susceptible to heavy infestations of the lime leaf aphid which produces abundant honeydew that can be a serious nuisance on pavements, cars and public seats as well as the building itself.
- Sycamore – this tree's leaf aphid also produces copious amounts of honeydew.
- Cedars – the soot aphid *Cedrobium laportei* can cause these trees to die back as well as produce honeydew. This will often become colonized by black 'sooty moulds' which add to the disfigurement.

Hard surfaces can also be stained by the fruit of trees such as cherries and rowans. The sticky buds of horse-chestnut, balsam poplar and general heavy leaf-falls can also have a detrimental effect. Many trees will also attract roosting birds who will defecate on any surface (and anyone!) below. Starlings can be particularly troublesome in this respect.

Although many 'clean' varieties of trees have been developed they still potentially provide a rich 'soup' of biological matter which can fuel the staining process.

3.3.2 Non-biological soiling

The atmosphere is not clean. It contains pollutants from natural sources such as pollen, bacteria and dust as well as 'air pollution' that occurs as a result of human activity. In relation to their impact on staining of buildings, the most important pollutants are divided into three groupings based on size (Hawes, 1986, p. 9):

- *aerosols or gaseous pollutants* – gaseous molecules;
- *soot or black smoke* – particles smaller than 1 μm;
- *coarse particulate matter or grit and dust* – particles greater than 1 μm.

A more complete list of pollutants is shown in Table 3.1.

Gaseous pollutants. Sulphur dioxide (SO_2) is the most active of the pollutant gases and is released as a result of the burning of fuels such as coal, coke and certain fuel oils. Its main corrosive effect is due to the formation of sulphurous and sulphuric acid that can dramatically break down a substrate. Approximately one fifth of the sulphur dioxide in the air is removed by solution in rain droplets, a similar amount is blown out to sea and the rest is dissolved by the water already on buildings, soil and vegetation. Surface condensation on the face of building materials can account for much of this moisture (Andrews, 1994, p. 13) and, as ambient humidities in some parts of the country can remain over 80% for much of the winter, the geographical location and the microclimate of a building can affect the rate of soiling. Lacy (1977, p. 15) has mapped average winter humidities across the whole country and revealed that higher relative humidities are found in coastal areas with lower levels inland especially in the lee of hills (see Figure 3.9a).

Rainfall is naturally acidic due to the presence of carbon dioxide (pH 5–6) but pollution makes it worse. Hawes (1986, p. 9) points out that the most acidic rainfall recorded in this country was in Scotland with a pH value of 2.4 – a greater acidity than the vinegar you put on your chips!

The corrosive effects of sulphur dioxide are not limited to damp environments. Evidence suggests that the gas can accumulate on dry surfaces, possibly by combining with other solid pollutants such as soot. Concentrations build up with the wetting/drying process and, when carried down façades or absorbed into materials, sulphates, chlorides and other substances may form which often cause the destruction of many materials.

Table 3.1 Classification of particles in the atmosphere (Carrié and Morel, 1975)

Diameter of particles (μm)	Types of particles
0.0001–0.001	Gaseous molecules
0.01–0.1	Smoke (tobacco, coal, fuel oil, metallurgical industry)
0.1–1	Smoke, bacteria
1–10	Bacteria, fog, mineral dust, fly ash (coal, fuel oil)
10–100	Dust from mineral, cement and iron and steel industries, fungi spores, pollen, mist
100–1000	Rain, silt, rock debris, dust, fly ash

Soot or black smoke. These terms refer to the visible products of imperfect combustion. The particles are small and behave like a gas in the atmosphere. In a diluted form, this smoke can form the 'bluish haze' that obscures distant objects in polluted environments. Smoke mainly consists of carbon or carboniferous matter which is nearly black. Coal can also contain tarry hydrocarbons that have an adhesive quality resulting in sooty deposits. These particles are supported by the air and can travel great distances from their source; when deposited on the face of a building they are not readily removed by flowing water. Soiling of this type can also prevent the evaporation of water from materials such as sandstone which influences durability. The particles block pore spaces, reduce permeability and restrict the movement of water both in and out of the material and as water is the most important factor in the breakdown of sandstone, decay rates may be accelerated (Andrews, 1994, p. 12). Accumulations of sooty deposits may be hygroscopic and able to hold large amounts of water, and so become very corrosive due to the inclusion of sulphur dioxide (Addleson and Rice, 1994, p. 144).

Grit and dust. These are particles that are large enough to fall under the influence of gravity and so are deposited close to the source of emission. They can include ash and unburnt fuel but also wind-blown dust from roads and industrial installations.

3.3.3 Efflorescence and lime staining

In the Preface of this book the scope of this study was clearly defined as not including what could loosely be described as 'defects' of buildings. Nor will it encompass temporary or easily removed blemishes or disfigurements. The exception to these rules must be the linked phenomena of efflorescence and lime staining. This is because they are so closely associated with the staining process that avoidance of them will also contribute to minimizing premature staining and vice versa.

Efflorescence. The mechanisms and causes of efflorescence have been fully described in many other publications (Addleson and Rice, 1994, p. 348) and will only be summarized here. It occurs when soluble salts in porous materials (bricks in particular) are deposited at or near the surface as the result of the evaporation of water that contain the salts. Progressively, as the water evaporates, the salts crystallize out to form harmless superficial white deposits on the surface of the building. It has been described as usually being 'a skin trouble and not a deep seated disease' (Addleson and Rice, 1994, p. 248). Apart from the few instances where efflorescence can be damaging to the sub-strata, it will be most familiar as that temporary disfigurement on new buildings that is normally removed by weathering agents within a year or so of construction. Even where it does return in 'cycles' following wetter weather, unless there is a supply of salts from another source the deposits will gradually reduce in intensity until it disappears altogether. In new work the extent is often related to the amount of rainwater that enters the structure during the construction process.

Post-construction efflorescence will always occur where percolating water affects the materials. Salts will be taken into solution and deposited on the edge of the damp-affected area.

Lime staining. Lime staining is similar to efflorescence and occurs when percolating water leaches free lime from cement and lime-based products, bringing it to an exposed surface where it is converted to an insoluble form of lime by exposure to carbon dioxide in the atmosphere. This is not temporary in the way efflorescence is and can be very difficult to remove once it has formed, often requiring chemical treatment (see Figures 3.6a and b). Avoiding lime staining is no easy matter as all cement-based products contain lime to greater or lesser extents. Lime is often incorporated in mortars used to bed down coping components and to ensure that the mortar is 'flexible' enough to avoid excessive cracking between the mortar and the coping unit. The only certain way of avoiding the problem is to control the flow of water through the construction by ensuring that water-excluding details are effective during the life of the building. In this way the avoidance of efflorescence, lime staining and premature staining is all linked through one element – controlling the flow of water.

Figure 3.6 Lime staining on (a) brick parapet, and (b) circular wall.

3.3.4 Changes in pollution levels

The nature of pollution levels in this country has changed significantly since the passing of the Clean Air Acts in 1956. Before this time the impact of pollution on buildings was often only incidentally documented. For example, at the turn of the century builders were given clear advice when choosing walling material (Rivingtons, 1904, p. 87). Two rules were offered: the first advised the designer to study buildings in the immediate neighbourhood while the second rule warned to

> be chary of adopting new materials . . . had Sir Gilbert Scott acted upon a rule like this he would not have specified Bath stone for the beautiful church at Haley Hill, in smoky Halifax, nor would the vicar and church wardens have been called upon to 'restore' the church within a few decades of its erection.

Nearly half a century later the streets of Walthamstow in north London were similarly affected. In the immediate post-war years

> coal was still king in the factory, around the hearth and on the railway . . . As a result the buildings were grimy black, especially the public ones – libraries, town halls and schools. The rich variety of British geology as reflected in our monumental masonry did not begin to be appreciated until the smog had gone and the high pressure water hose had arrived in the later 1960s.
> (*Hennesy, 1992, p. 88*)

The introduction of the Clean Air Acts in 1956 removed the predominance of coal-based pollutants but produced few interesting changes in the nature of pollution in this country:

- The amount of smoke in the atmosphere has reduced allowing more of the sun's heat through the air. This has increased the dispersion of sulphur dioxide to the upper levels of the atmosphere where winds carry it further afield and eventually down to the ground. This 'acid rain' has now become a problem for countries that have never before been affected by pollution.
- Urban concentration of smoke has fallen by as much as 70% but is still about twice as much as rural areas.
- Analysis of pollution monitoring in city centres suggests that as much as 77% of particulate matter now comes from vehicles.
- Smokeless zones have had less impact on sulphur dioxide than on smoke emissions.

The smogs may have gone but atmospheric pollution still has a significant role in the staining process.

3.4 THE SURFACES OF MATERIALS

The staining of façades will depend on a number of factors:

- the nature of the material surfaces;
- the amount and type of staining agents;
- the influence of the wind and rain.

The type of material is not only important but also the characteristics of its surface, as this will affect the adhesion and the visibility of any surface contaminants. The attributes of a material's surface may be quite different than those within the material itself. The surface of some materials (e.g. concrete) may also change dramatically during the initial period of exposure and other materials may develop a new skin after more prolonged weathering.

Façade materials can be categorized according to their different porosity values:

- *Porous materials* – concrete; bricks; natural stone and wood.
- *Non-porous materials* – glass; plastics and metals.

An understanding of the nature of the surface of materials is not well developed and does not always give a clear insight on how materials will weather and stain. A more thorough consideration of this aspect is considered by Verhoff (1988, ch. 1).

Addleson and Rice (1994, p. 566) examine broader influences on the way a material will be affected by staining and influence the flow of water over it. This is summarised below.

3.4.1 Absorption and absorptivity

This relates to the ability of the material to take in liquids (i.e. water):

- *Absorption* – relates to maximum amount of water that the material can absorb.
- *Absorptivity* – is the rate at which the material can absorb water.

These are both controlled by the porosity of the material. These properties are crucial as they determine the incidence of surface flow of water over a material and so affect the staining process.

3.4.2 Dirt retention

'Dirt' can be regarded as a general term that relates to the wide range of particles that may cause staining. The amount of particles retained will depend on:

- their size
- their adhesive properties

It is likely that porous materials with an open pore structure will hold more dirt than non-porous ones. There are two types of adhesion:

- Mechanical adhesion occurs when the dirt particle locks into the pore structure.
- Specific adhesion will take place between two surfaces where one can 'wet' the other and the strength of this adhesion increases with physical wetness.

3.4.3 Texture

Materials that have a strongly textured surface pose the designer with the ultimate dilemma in relation to staining:

- Textured surfaces break up and disperse the water flow over the surface and reduce the contrast between

washed and unwashed areas. They can also disguise staining.
- Textured surfaces also retain dirt so that in overall terms the amount of particles can be greater than for a smoother material.

3.4.4 Colour

Observations of buildings reveal that light-coloured materials show up staining more clearly than darker ones. Portland stone is the classic example that shows a strong contrast between washed and unwashed areas.

3.4.5 Solubility

For some materials, flowing water can dissolve constituents from it and deposit them on another adjacent material. The most notable examples include water flowing off concrete on to brickwork allowing deposition of white blotches of calcium carbonate and the run-off from copper components producing green staining on other materials. A typical example of this is shown in Figures 3.7 and 3.8.

Addleson and Rice (1994, p. 568) usefully describe some of the weathering/staining characteristics of common façade materials and state: 'most of this information is qualitative simply because the relevant quantitative data do not exist'.

3.5 INFLUENCES OF THE WIND AND THE RAIN

3.5.1 Prevailing wind and driving rain

Water has a profound effect on the performance of buildings. It is involved in a number of destructive mechanisms (Robinson and Baker, 1975, p. 2):

- dimensional change;
- corrosion;
- leaching;
- efflorescence;
- water penetration leading to deterioration of internal finishes.

Figure 3.7 Rear elevation of a bus shelter in London that has been faced with smooth stone cladding panels. The roof was metal with a copper component.

Figure 3.8 Detail of roof/cladding junction. Run-off from the roof components was allowed to flow over the stone depositing green/grey stains on the face.

In relation to the premature staining of new buildings the action of rain has a central role as it facilitates both biological and non-biological soiling. Because most rainfalls are associated with winds, a building's exposure to the prevailing wind becomes increasingly important. Driving rain, which is defined by Robinson and Baker (1975, p. 2) as 'rain carried along by wind at an angle to the vertical', will vary directionally over time. For Great Britain these have been assessed by Lacy (1977, p. 109) and, through a series of nationwide compass roses that show the percentage of driving rain for each of eight compass directions, are shown allowing general comparisons of the exposure for different locations (Figure 3.9). These must be adjusted to take into account local variations such as altitude, shelter and other special conditions related to the specific site, so it tends to reveal annual summaries of conditions rather than the great variations that will inevitably occur in any one place.

This information is useful to designers so building components can be selected to suit the worst exposure conditions for a particular site but is not refined sufficiently to enable levels of staining to be predicted. This is because:

- staining is a very complex process where the incidence of driving rain is just one element; and

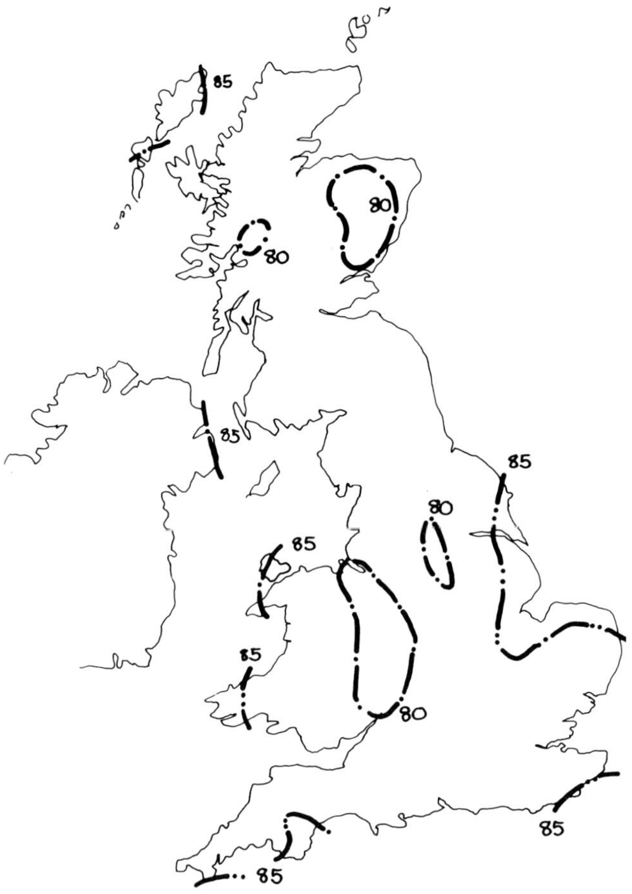

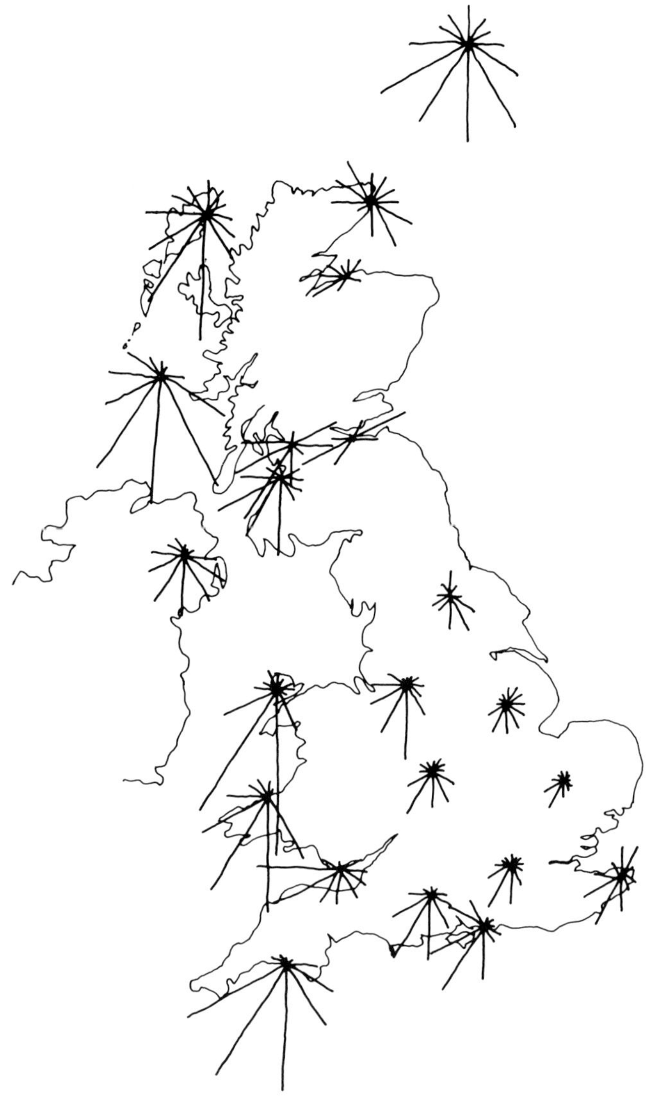

Figure 3.9 (a) Average means of relative humidity in the British Isles during January; (b) map of the UK showing annual mean driving rain roses, 1956–66 (Lacy, 1977).

- localized conditions (i.e. shelter/exposure provided by adjacent buildings) will vary wind patterns significantly even across the same elevation.

Despite this, prevailing wind conditions can give an indication from which direction most of the rain will come and on many tall buildings this will result in greater levels of washing on the exposed elevations while soiling will be dominant on sheltered sides. The buildings in Figures 3.10 and 3.11 clearly show this effect.

3.5.2 Rain against buildings

Although dominant wind direction can provide useful background information, the knowledge of how rain strikes an external wall and behaves afterwards gives an insight into how water redistributes dirt on a façade. Beijer (1980) studied this and although he was concerned with concrete, the principles can be applied to vertical surfaces of any material. This work found that the path of raindrops is influenced by:

- the force of gravity;
- mass forces;
- friction forces acting on the air as it flows around surfaces and obstacles.

Because wind force, direction and rain intensity vary greatly even within the same rainfall, it is not possible to produce any accurate calculations. Beijer (1980, p. 13) highlighted a few revealing characteristics of rain:

- The average vertical rate of fall for rain is about 4–5 metres per second (m/s) which remains constant for most types of rain.
- The diameter of raindrops is estimated to be between 2–5 mm which also remains the same between one rainfall and another.

Combining these assumptions with information on air currents, directions and velocities, the approximate paths that raindrops might take around a 20 m high building were calculated and are illustrated in Figure 3.12 (Beijer, 1980, p. 14). The main outcomes of this study were that:

- Less than half the quantity of rain that should pass through an equivalent cross-sectional area of 'free air' is caught by the external wall. This applies regardless of the wind force.

Figure 3.10 Photograph (a) shows an ornate stone façade that faces the Mersey estuary at Liverpool and is almost completely washed clean by the driving rain. In (b), the leeward elevation is heavily stained and is disfigured by differential washing.

Figure 3.11 The elevation to the right of this public library faces south west while that to the left of the photograph faces north east and is heavily soiled.

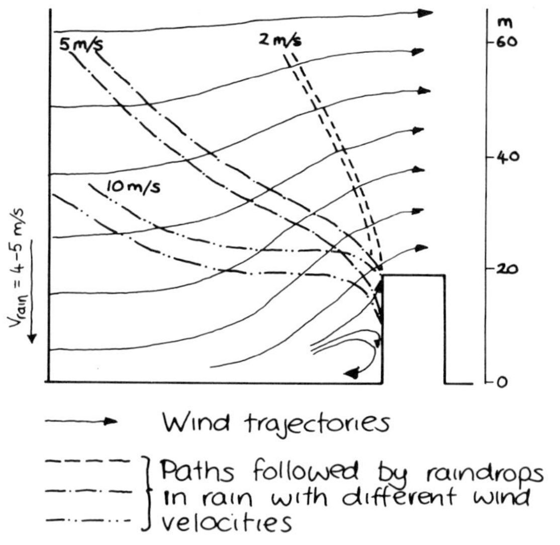

Figure 3.12 The path followed by raindrops against a tall building approaching perpendicularly at different velocities (Beijer, 1980, p. 14).

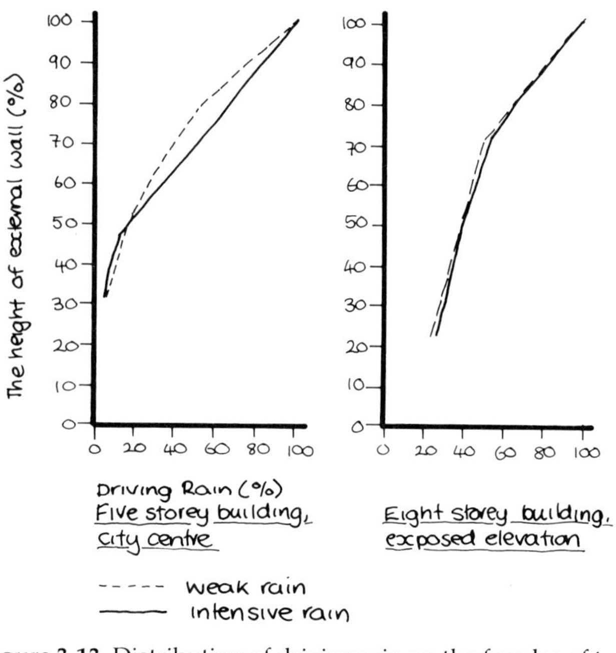

Figure 3.13 Distribution of driving rain on the façades of two different buildings. The comparison of weak and intensive rain for a sheltered and free-standing building illustrates there is little difference in the distribution between the two.

- The rain strikes mainly the top parts of the external wall.
- Raindrops move almost parallel to the lower sections of the external wall.

Figure 3.13 shows the distribution of driving rain on the vertical face of two buildings for both intensive and weak rainfalls and demonstrates that there is little difference especially at higher levels.

This was confirmed by Robinson and Baker (1975, Fig. 11). Where airflow abruptly changes direction around the edges of a building it is likely that the raindrops are unable to follow and so strike the face of the building. This would account for the wetting of windward corners, parapets, protruding cornices and projections such as columns and cills. Figure 3.14 shows a typical wetting distribution for a tall building.

Research into the deposition of rain on and around projections from the face of a façade has produced some interesting results. Beijer (1980, p. 18) showed that even fairly large projections (up to 1 m projection from a vertical face) can be washed clean by driving rain. Figure 3.15 illustrates this effect and could result in large features such as projecting bays being washed clean while adjacent wall surfaces remain dirty.

While observing weather protective features on an experimental external wall, Herbert (1974) found that projections on the lower region of a test façade protected the wall below them while those projections in the upper area resulted in considerable concentrations of driving rain directly beneath the projection itself. This was far in excess of what would have struck an equivalent plain wall. Although Herbert pointed out that overhangs

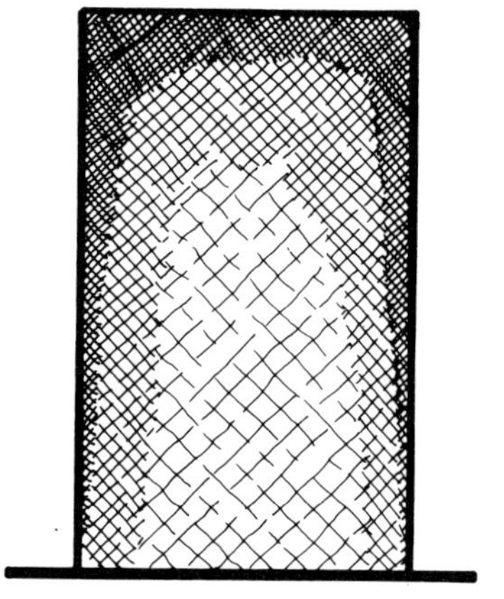

Figure 3.14 Typical wetting pattern on the face of a multi-storey building subjected to wind-driven rain.

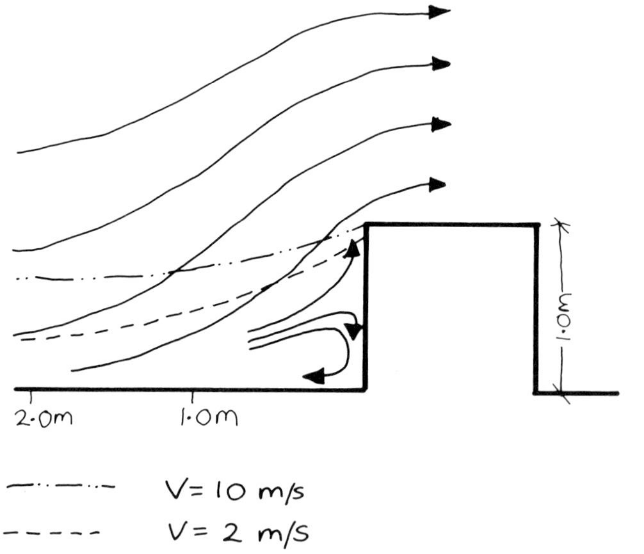

Figure 3.15 Change in direction of raindrops travelling horizontally when approaching an obstacle projecting 1 m from the face of a vertical wall. Numbers represent wind velocity zones of air when it meets the side of the obstacle (Beijer, 1980).

provide considerable protection against water run-off, these high concentrations of water may account for walling failures on tall buildings that were apparently well protected by overhanging eaves. It is interesting to note that the smaller the overhang the higher the level of water collected. This could be due to an 'air cushioning' effect below larger projections preventing high levels of deposition.

Although this study was carried out on a specialized test rig in an exposed position using metal overhangs, the results must question the effectiveness of the traditional cornice/overhang on tall buildings proposed by many commentators as an essential protection against premature staining. Diagrammatic details of this phenomena are shown in Figure 3.16.

A theoretical method of calculating the amount of shelter provided by projections was investigated by Ishikawa (1977). This study attempted to calculate the area of soiling that would occur because of the sheltering effect of deep roof eaves, balconies and sill projections. Mathematical methods were formulated but it was accepted that the unpredictable nature of driving rain would introduce so many variables that the application of this approach would be limited in practice.

3.5.3 Movement of water on the face of a building

Once water has been deposited on the face of a building its movement across the façade will determine whether the component will stain or be washed clean. The characteristics of rainwater run-off will be largely determined by the capillary suction capacity and moisture content of the wall material. For example, brickwork and render will absorb almost all deposited rain while metal and glazed walls will take up little, allowing rain run-off to develop quickly. The basic factors involved in this were expressed by Addleson and Rice (1994, p. 563) where the rate of flow over a surface (w_r) are the quantities of incident water (w_i), the height of the surface (h) and its absorptivity. The rate at which the rain hits the surface is governed by its exposure and the absorptivity of the material controls the proportion of the water absorbed (w_a) and the proportion that flows over the surface (w_r). The main restriction of this model is that it will depend on the intensity of the rain which varies tremendously even within the same rainstorm (Figure 3.17).

Beijer and Johannson (1976) measured and calculated the magnitude of rain water run-off on a 20 m high flat concrete wall. This revealed that it took 18 minutes for a run-off to appear which gradually increased as the façade material absorbed moisture. Even though the rainfall was quite intense, the run-off never reached ground level. Figure 3.18 shows this effect in which the curves

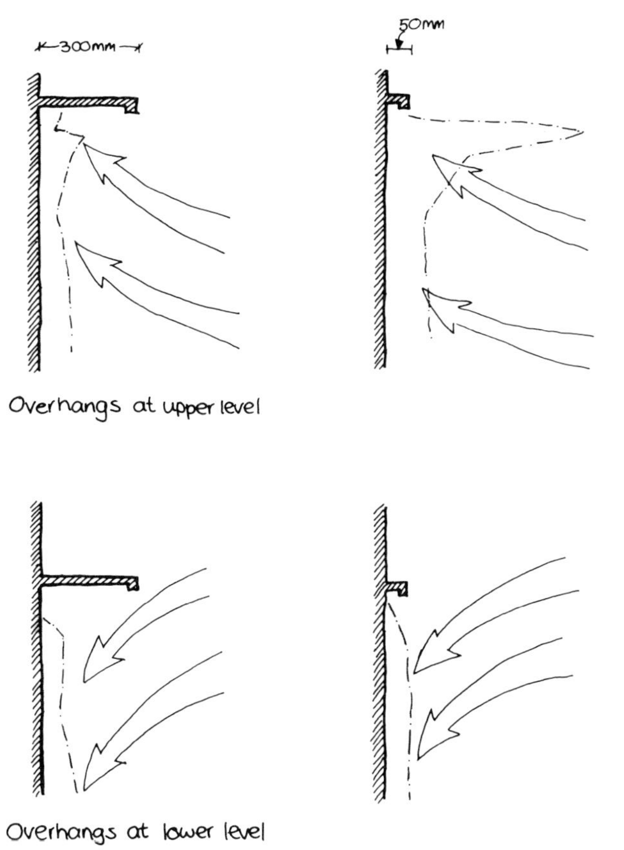

Figure 3.16 Diagrammatic representation of the amount of water collected below metal overhangs on a test façade rig at varying wind speeds (Herbert, 1974).

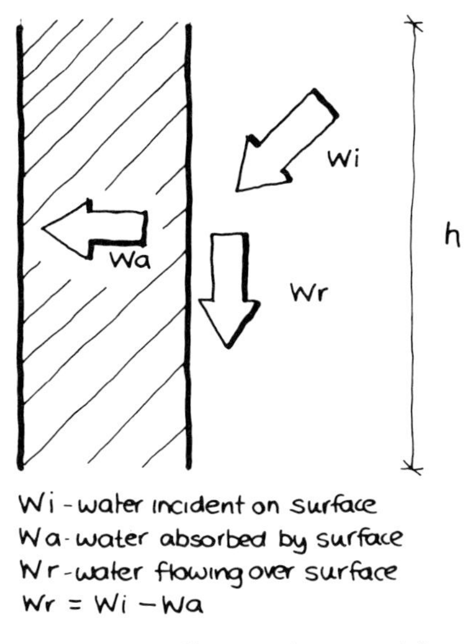

Figure 3.17 The factors affecting the rate of flow over a wall surface (Addleson, 1994).

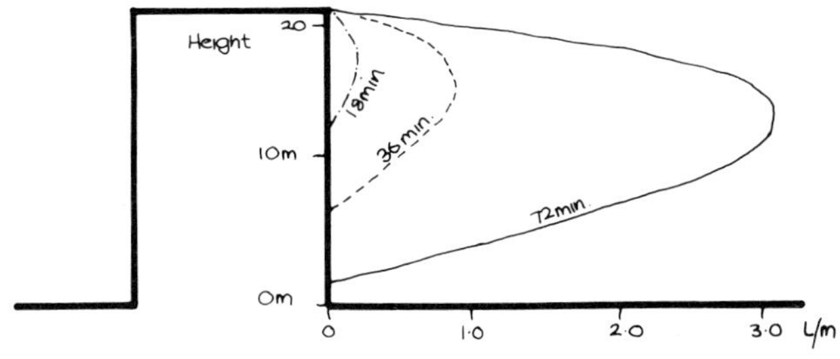

Figure 3.18 Calculated run-off streams on a 20 m high flat external wall surface of concrete. This assumes medium intensity rain of 1 mm per hour against the top part of the wall. This clearly shows that even after 72 minutes, the run-off never reaches ground level.

represent the duration of the rainstorm and the total quantity of water flowing past a certain point is shown on the horizontal axis.

Observations of rain run-off down a 14-storey building by Robinson and Baker (1975) produced similar conclusions. Here rain run-off did not begin until 20 minutes into the rainfall and when the shower had finished 40 minutes later the run-off had only extended three storeys down the building.

Further work by Beijer and Johannson (1976) looked at a whole range of façade materials including 'standard' brickwork, render, hard-burnt brick, limestone, concrete, metal and glass. Based on Swedish climates, the study found that for average conditions water run-off on taller buildings constructed of brick and render rarely reached the ground, for limestone façades the incidence was higher, with concrete representing the highest probability. In the case of metal and glazed façades where absorption is virtually negligible, run-off down to ground level occurred with only weak rainfalls limited only by the façades' ability to retain individual raindrops on their surfaces.

Cronshaw (1971) carried out a pilot survey to study the relationship between the amount of water striking the face of a wall as driving rain and the amount running down the surface. Based on the collection of water run-off from four different façades and comparing the results to general rainfall levels, the following conclusions were drawn:

- For walls of moderate exposure, the run-off down the face of the wall is considerably less than the volume of driving rain striking the wall even when the wall is saturated.
- The proportion of run-off compared to impinging rain increases with the length of the rainfall but rarely exceeded one twelfth of the total striking the wall.
- The amount of run-off varied greatly depending on the nature of the surface, its characteristics and position on the building.

Beijer's important work identified other important rules governing water flow down a façade:

- Rain run-off on many surfaces is a very thin layer, probably no more than a few tenths of a millimetre, and flows at very low velocities, no more than 1 m per minute.
- Run-off is mainly vertical with lateral winds having little effect on its distribution over porous surfaces.
- On smooth flat surfaces run-off tends to break up into separate streams that tend to follow consistent paths.
- On vertically projecting features, run-off from driving rain often forms concentrated streams of clean water below the feature that washes the surface. Run-off will also follow vertical joints causing a similar washing effect where the joint terminates.
- Horizontal projections on façades often give rise to moderate levels of run-off of very dirty water which is often quickly absorbed after flowing only a short distance below. Horizontal obstructions also tend to spread out the run-off from vertical surfaces above to form less concentrated water flow with possible different soiling actions. The design of drips is also very important because of the thinness of the water flow and will regularly flow around the underside of a projection unless it has an effective throating.

3.6 CONCLUSIONS

This brief review of the mechanisms of staining has revealed that the process is very complex and cannot be accurately predicted. General characteristics of materials and influences of structural form can be identified but when combined with the actions of wind and rain the variability becomes overwhelming. Despite this, designers must be familiar with these processes so that the worst excesses can be avoided.

REFERENCES

Addleson, L. and Rice, C. (1994) *Performance of Materials of Buildings*, Butterworth–Heinemann, Oxford.

Andrews, A. (1994) *Stone Cleaning. A Guide for Practitioners*, Historic Scotland, Edinburgh.

Beijer, O. (1980) *Weathering on External Walls of Concrete*, Swedish Concrete Research Council, Swedish Cement and Concrete Research Institute, Stockholm.

Beijer, O. and Johannson, A. (1976) *Driving Rain Against External Walls of Concrete*, CBI Research 7: 76, Stockholm.

Carrié, C. and Morel, D. (1975) *Salissures de façades*, Editions Eyrolles, Paris.

Cronshaw, J.L. (1971) Rainwater run-off from walls. *Building Magazine*, 12 February.

Hawes, F. (1986) *The Weathering of Concrete Buildings*, Cement and Concrete Association, Slough, p. 2.

Hennesy, P. (1992) *Never Again, Britain 1945–1951*, Jonathan Cape, London.

Herbert, M.R.M. (1974) *Some Observations on the Behaviour of Weather Protective Features on External Walls*, Current Paper 81/74, Department of Environment, BRE, Garston.

Hibberd, B.G. (1989) *Urban Forestry Practice*, Forestry Commission Handbook 5, HMSO, London.

Ishikawa, H. (1977) The extent of shelter provided by projections on external walls from driving rain. *Evaluation of Building Performance: External Wall Surfaces*, RILEM, Espoo, Finland.

Lacy, R.E. (1977) *Climate and Building in Britain*, Department of Environment, BRE, Garston.

Rivingtons (1904) *Rivingtons Series on Building Construction*, Longman Green and Co.

Robinson, G. and Baker, M.C. (1975) *Wind Driven Rain and Buildings*, Technical Paper 445, National Research Council of Canada, Division of Building Research, Ottawa.

Verhoff, L.G.W. (ed.) (1988) *Soiling and Cleaning of Building Façades*, Report of the Technical Committee 62 SCF, RILEM, Chapman and Hall, London, p. 61.

Chapter 4
Premature staining: case studies

4.1 INTRODUCTION

The previous chapters aimed to introduce the context and the mechanisms of premature staining of buildings. It was not a thorough study but one designed to give an overview of the work done in this area. Another reason for only looking at the broad principles is that not much research exists on this topic. According to Hawes (1986, p.12), 'we have no data, apart from casual observation, related to the action of water on the more sheltered parts of buildings because all known work has been principally with rain penetration'. This is indeed true; few researchers have been concerned with the aesthetic effects of the wind and rain, but instead have concentrated on the more fundamental aspects of weather exclusion. Where staining has been studied it has been in relation to concrete structures – a throwback to the 'brutalism' of the 1960s and 1970s when large-scale concrete buildings weathered very badly. Until recently, the only comprehensive publication to look at the mechanisms and effects of staining has been by the Technical Committee of RILEM (the International Union of Testing and Research Laboratories for Materials and Structures – Verhoff, 1988). Although an important publication, its highly technical content, style, mode of presentation and marketing have not made its valuable advice and guidance accessible to designers.

A more recent publication that has dealt with the disfigurement of natural stone buildings was written by the Masonry Conservation Research Group at Heriot-Watt University (Andrews, 1994). Although this was primarily concerned with the cleaning of sandstone buildings in Scotland, the study did take a broad view of the problem including conceptual understandings of the staining process (which is explained in Chapter 5).

4.2 METHODOLOGY

Resources did not allow for 'controlled studies to verify observations in the field' (Hawes, 1986, p.12) but instead this chapter is based on the visual observations of a number of recently completed buildings in this country. No special tests were carried out or inspection routine followed. Some observations were made casually and quickly while other examples have been monitored over a number of years. Existing research has shown that staining prediction models are not reliable because of the large amount of variables involved in the staining process, therefore the monitoring and recording of the performance of real buildings in their own climatic environment is arguably just as good as any other scientifically sound methodologies.

4.3 STRUCTURE OF THE CASE STUDIES

The case studies reflect the type and incidence of staining on buildings visited rather than a systematic investigation of how different elements of buildings perform. Accordingly, some parts of buildings are adequately covered while others may receive shorter commentaries (if any). After visiting several hundred buildings it is hoped that the main causes of staining have been included. In each category a short reflective section has been included that looks at 'best practice' and in some cases presents new and novel ways of overcoming some common staining problems that designers (hopefully) may find useful.

4.4 COPINGS

This element has been included first and in most depth as it is probably responsible for the vast majority of disfigurements on new buildings.

4.4.1 Standards

The design and specification of copings is covered by British Standard 5642: Part 2 (BSI, 1983) where a coping is defined as a 'unit placed at the head of a wall and designed to shed water clear of all exposed faces of the walling it is intended to protect'. Copings are the most exposed parts of a wall, often being wetted from two sides as well as from rain travelling downwards. On taller buildings, localized winds travelling vertically and parallel to the surface of the façade can dramatically reduce the coping's ability to shed water. This level of exposure often results in free-standing parapet walls remaining wet for most of the winter, creating ideal conditions for damage from frost, sulphate attack, mechanical damage and organic growth. It is no surprise that these features are the first (and often worst) cases of premature staining.

The different types of coping have been categorized as follows:

- *Brick copings*
  - copings and walling in same quality brick (porous and non-porous);
  - copings in impervious brick, walling in porous brick;
  - copings with tile creasing.
- *Pre-formed coping units*
  - clay
  - concrete
  - natural stone
  - metal.

Copings included in this study were found in all locations on buildings such as free-standing boundary walls; parapet walls around flat and pitched roofs; balconies and roof gardens; walls to external staircases and planting areas.

4.4.2 General aspects

From a staining point of view, copings present the designer with a number of challenges:

- They tend to collect more water than adjacent wall areas.
- Because of their flat or slightly inclined surface, a high level of pollutants can accumulate very quickly.
- They provide excellent roosting places for birds that defecate on them.
- The method of discharging dirty run-off to lower levels is both crucial and difficult to predict and control.

The different categories will be considered in more detail.

4.4.3 Brick copings

Coping and walling in the same brick (porous). Whatever the location, the exposure rating of a wall will vary with height, the most exposed being the coping itself. Therefore if masonry units are selected to match these changing conditions the brick type should ultimately vary. Yet many designers use bricks of the same grade throughout the wall, often for aesthetic reasons rather than their durability. The most concentrated disfigurement will be on the top of the coping itself but as water is shed the visual impact will extend down the vertical surfaces. Dirt accumulation, algae, moss and lichen can quickly disfigure the wall surface.

The extent to which the staining will extend down the wall will depend on:

- porosity of the wall
- exposure of the site
- rainfall level.

Where the mortar joints are recessed, moss growth and other higher plant forms are more likely to take hold (Figure 4.2).

Figure 4.1 General view of brick parapet wall, coping and walling in the same porous brick. This shows algae staining concentrated on the coping but extending down the wall face.

Figure 4.2 Brick coping with relatively impervious bricks and porous mortar. Algae and moss growth is quickly followed by grass and other higher-order plants. Tile creasing is also shown that has been laid flat without weathered mortar pointing allowing moss and lichens to become established.

Figure 4.3 General view of a boundary wall where coping and walling are in the same quality non-porous brick.

Copings and walling in same brick (non-porous). In this case, the bricks usually remain relatively clean with the staining concentrating on the mortar joints. How obvious this staining often depends on the colours of both the units and mortar (Figures 4.3 and 4.4).

Coping non-porous and walling porous. Proportionally less moisture may be retained by the coping and so more water flows down the face of the wall extending the area of staining beyond that likely to result with a more porous coping (Figures 4.5 and 4.6).

Coping with tile creasing. A tile creasing is a traditional method of weathering a coping and throwing water clear of the wall beneath but in reality these features can create a number of problems:

- Creasings are often set level, allowing water to flow back into the wall increasing frost and/or sulphate damage.

Figure 4.4 Detail of Figure 4.3 showing the concentration of staining to mortar joints through algae and moss growth. Because the mortar is more porous than the bricks, its surface remains damp and provides a good growing medium.

Figure 4.5 A boundary wall where the coping bricks are non-porous and the walling units are porous. The staining is focused on the joints of the coping and the upper courses of the porous walling. Note that the inclined coping has concentrated the run-off to extend the disfigurement beyond 'normal' levels.

Figure 4.6 Engineering bricks have been used as copings over more porous facings. Staining has occurred at the joints and on the porous brick face. Run-off has been concentrated by the vertical mortar joints of the coping and the corresponding joints of the courses below, causing 'lines' of staining.

- The ledge created by the tile creasing is frequently pointed up with a weathered mortar bed. This will retain moisture allowing algae, mosses and lichens to grow (see Figure 4.2).
- Although the creasing does project beyond the face of the wall, without a formal throating on the underside, weather will flow haphazardly with some running back to the face of the wall. This flow is often guided and concentrated by the joints between the tiles.

The housing scheme described in the Preface of this book included a tile creasing feature which had been cut back flush to the face of the wall. Although this might have given a reference to a 'vernacular' style, it served no useful function (see Figures 0.1 to 0.4). The boundary wall in Figures 4.7 and 4.8 illustrates how a tile creasing below can actually cause staining rather than prevent it.

Figure 4.8 Detail of coping illustrated in Figure 4.7.

4.4.4 Pre-formed copings

Pre-formed coping units and systems have been regularly used by building designers to protect the top of walls and can minimize staining usually associated with brick copings. Despite this, it has been observed that even coping units manufactured to British Standard guidance do not prevent staining completely. One of the most notable visual effects are areas of concentrated disfigurement at the joints between the coping units themselves. Figures 4.9 and 4.10 show a very common form of staining, at the joints of a clay coping bedded and jointed in mortar.

Figure 4.7 The inclined boundary wall with a non-porous coping directs coping run-off directly on to the tile creasing. This flows down the slope and because of the lack of an effective throating beneath the creasing there is a concentrated water flow and consequent staining at the change in direction.

This concentrated staining at coping joints is probably caused by increased water flow through three different mechanisms:

- Thermal and moisture movements of the coping units aided by the slip plane of the dpc below can result in small cracks opening up between the coping units and the mortar joints; the larger the coping unit the greater the amount of possible movement. Water will enter the joint and flow down through a combination of capillary action and the influence of gravity. Figures 4.11 and 4.12 show this effect both graphically and photographically.
- The surface of the joint material (usually mortar) is different in texture and porosity than the adjacent coping units and is usually recessed, both factors possibly encouraging greater water flow down the joint when compared to the adjacent plain coping surface (Figure 4.13).
- Most copings are bedded and pointed with mortar on site. Through inadequate tolerances and poor workmanship this can result in the continuity of any throatings beneath the coping overhang being disturbed. This will allow water to flow back to the face of the wall (Figure 4.14).

Through a combination of these mechanisms, staining at coping joints can appear very early in a building's life. The initial disfigurement is usually caused by 'lime staining' where the flow will deposit calcium carbonate over the wall surface. Once this has ended (because it will be limited in duration) the conventional staining agents will become dominant.

Figures 4.9 and 4.10 These clay coping units have steeply sloping that have throatings cut on the underside. The profile sheds water quickly but concentrates flow at the junctions causing dirt staining, bacteria and algae growth that leave prominent streaks on the wall beneath.

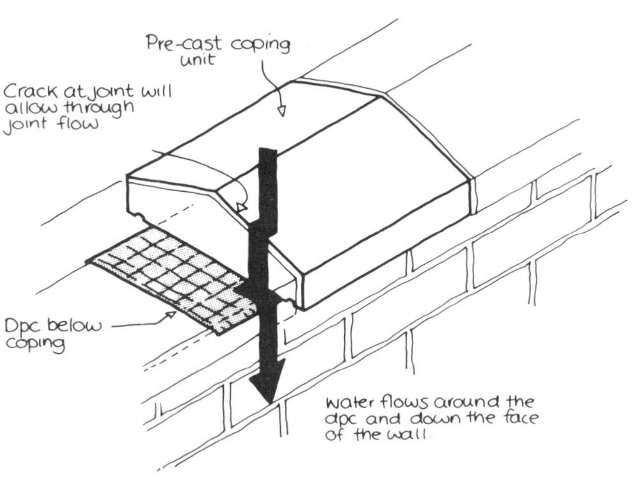

Figure 4.11 Diagrammatic representation of through-joint between adjacent pre-cast coping units.

Figure 4.12 Typical example of joint staining that is almost certainly caused by water passing through a crack between the mortar/coping interface.

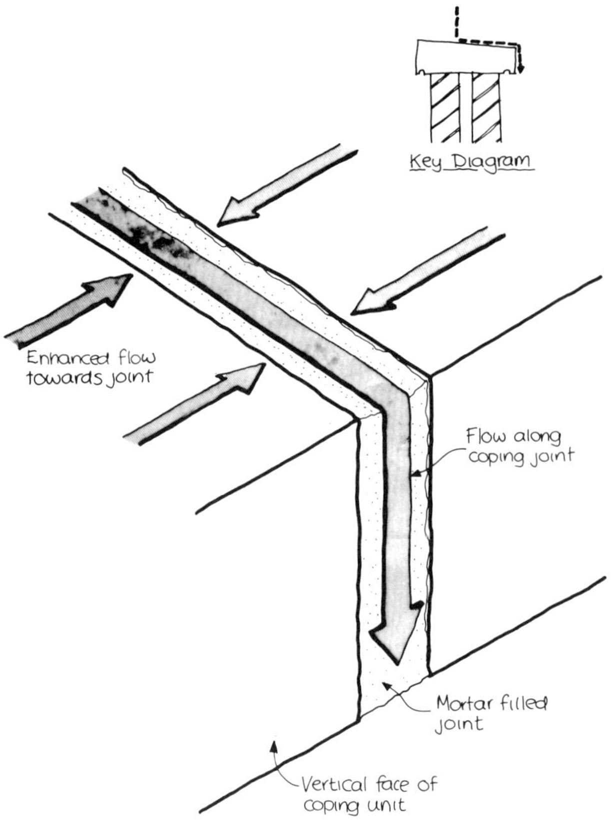

Figure 4.13 Increased water flow along joints between coping units.

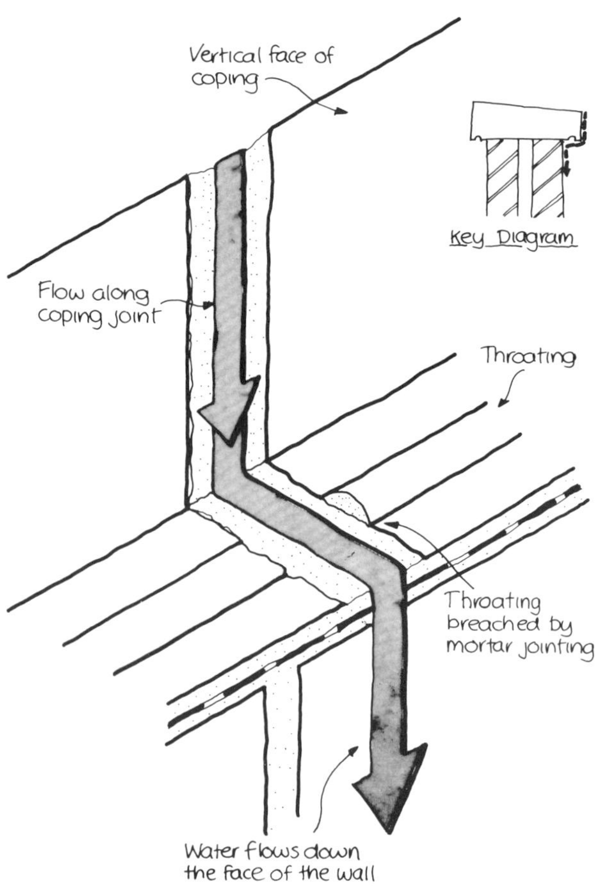

Figure 4.14 Diagrammatic representation of a blocked throating to the underside of a pre-cast coping system.

Copings on inclined or sloping parapets. Copings on inclined walls are often located alongside pitched roofs, external staircases and other architectural features. Many terminate with a horizontal or a springer section representing an abrupt change in inclination. A large amount of water can flow down the slope accumulating as it progresses and can be diverted down the face of the wall at joints with the most concentrated flow at the change in inclination. Throatings are often ineffective in preventing this. Figures 4.15 and 4.16 illustrate this effect. The building in Figure 4.17 is approximately two years old and shows both premature staining and lime staining at these identified locations. Staining is often worse where the junction between the final sloping unit and the first horizontal section is made by cutting and jointing on site. Where this change in inclination is formed out of one specially manufactured unit, staining is minimal if present at all. The design of the throating is crucial in this respect (Figure 4.18).

4.4.5 Pre-formed metal copings

Aluminium and other metal copings systems have been used for a number of years and usually consist of a coping section that clips over brackets fixed to walling below. The joints of the coping sheets are often waterproofed with mastic sealing strips or gaskets below. As no water is absorbed, high levels of run-off can occur and if the slope of the coping is minimal, a significant proportion of water can remain on the coping surface and will be 'sucked' into the joint through capillary action. This can flow down at the face of the wall causing staining (Figure 4.19).

A similar problem can occur with aluminium trims that are used at the verges of flat roofs where asphalt or felt coverings are terminated. Differential expansion can result in the tearing or splitting of the roof covering at the joint. If the slope of the verge is minimal a surprising amount of water can run down the face of the wall (Figures 4.20 and 4.21).

4.5 DEVELOPMENTS IN COPING DESIGN

Developments in design and manufacture have resulted in many improvements in the performance of modern coping systems although many of these have focused on stability, resistance to water penetration and vandalism – few initiatives have been targeted at improving the water shedding characteristics. The two major improvements have included:

- copings that fit and overhang the top of the parapet wall to allow better positioning of the coping and dpc (referred to as 'clip type' copings in BS 5642: Part 2, 1983);
- many systems also incorporate restraining elements that fit into grooves and recesses underneath the coping units which 'lock' them in position.

Figure 4.22 illustrates such a unit as manufactured by Nottingham Brick under its 'Topknott' product range.

Figure 4.15 Through-joint flow between coping units on an inclined wall. Although the flow will occur at the joints the staining will often appear further down the walling.

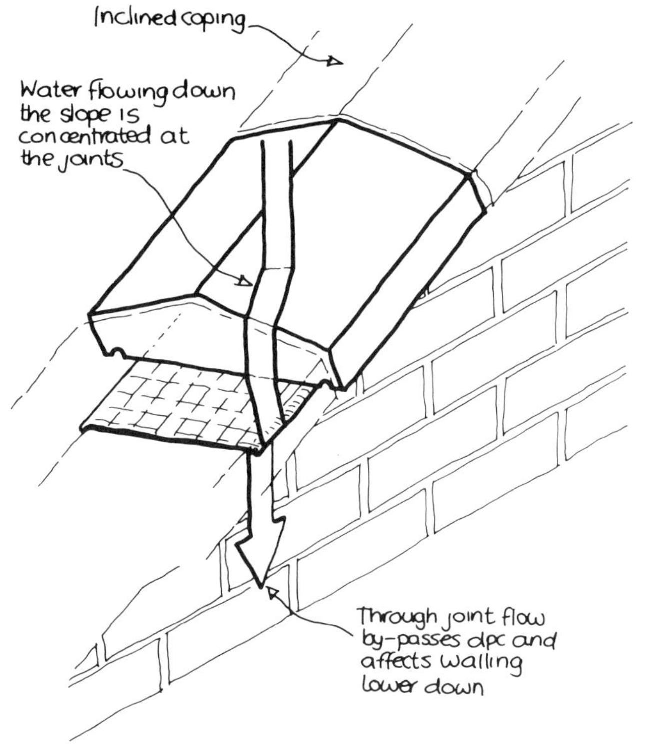

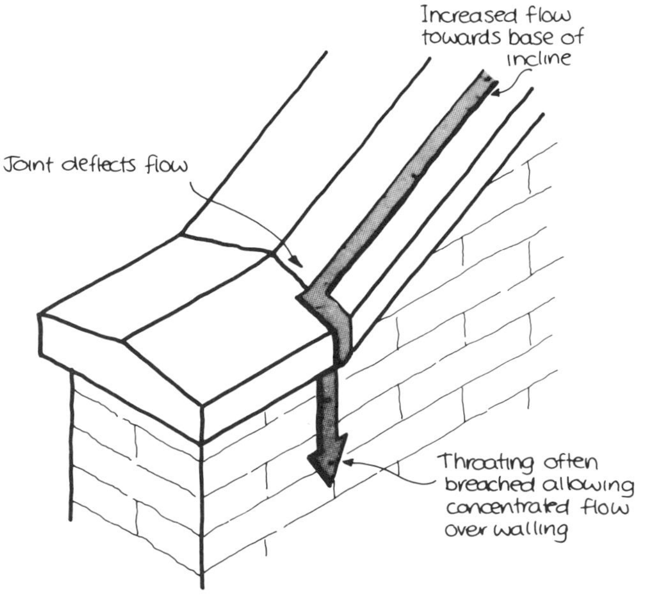

Figure 4.16 Water flow along inclined copings. Even if the top surface of the coping unit is sloped, large amounts of water can still run along to be discharged down the face of the walling at the change in inclination.

Figure 4.17 Staining caused by inclined copings. This two-year-old feature shows both the disfigurement of lime and premature staining.

Figure 4.18 The change of coping inclination on this building is bridged by a specially manufactured unit that keeps the throating entire thus avoiding staining.

4.5.1 Lessons from traditional practice

Traditional building practices, especially those related to the use of stone, have developed useful methods and techniques that, if adapted, may help produce coping designs that minimize premature staining problems. These and other key issues have been amplified in the following section.

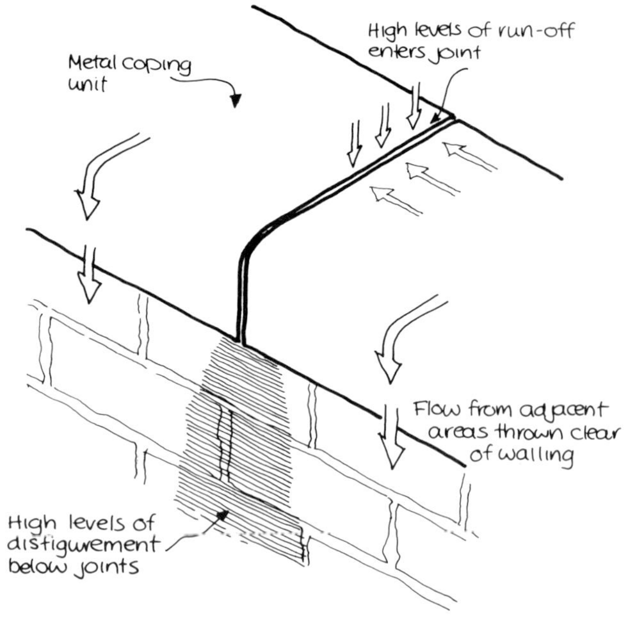

Figure 4.19 Staining at the joints of metal coping units.

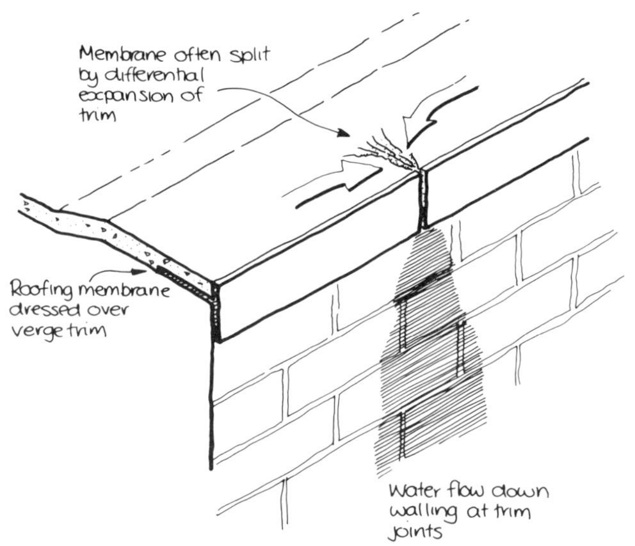

Figure 4.20 Water flow at the junction of verge trims to flat roofs.

Figure 4.21 Extensive water flow at verge trim joints caused considerable staining before additional drainage was provided to the flat roof of this building.

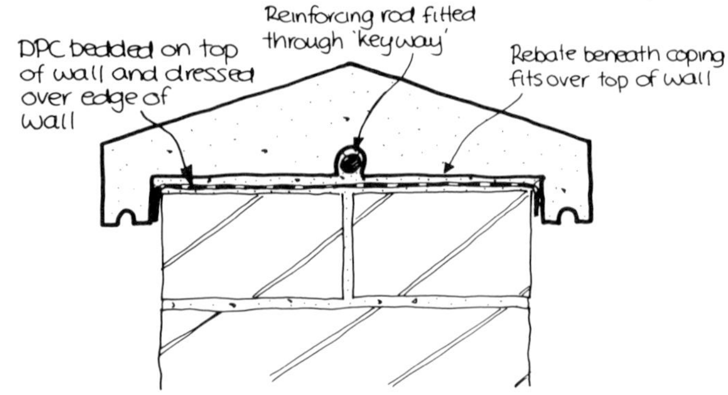

Figure 4.22 Nottingham Brick's 'Topknott' range of clip type copings.

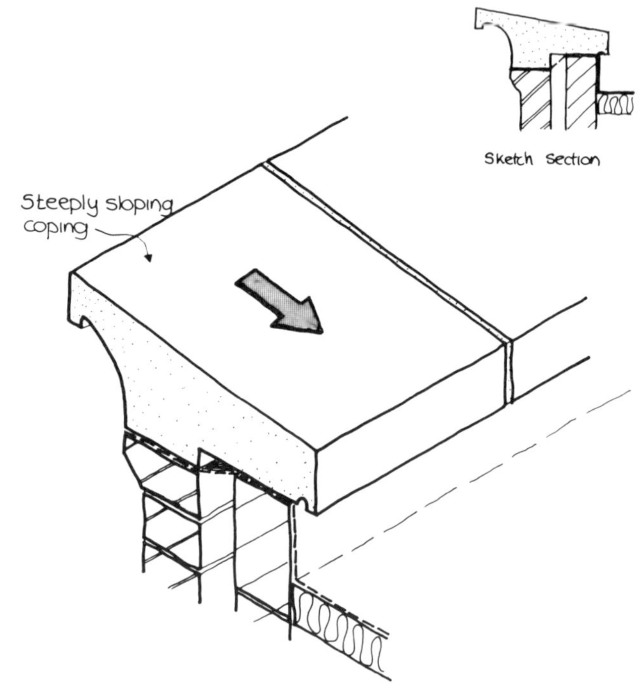

Figure 4.23 Coping detail from the Truro Courts of Justice showing the use of steeply sloping surfaces.

4.5.2 Key issues in coping design

Slope of the surface. The slope or fall of the coping is crucial. Where the gradient is shallow, rainfall will result in water flow over the face of the coping and down the wall surface. Therefore slopes on copings should be maximized where possible. The court houses in Truro that won an AJ award (Dunster, 1988) used specially designed copings that incorporated steeply sloping surfaces. Although the performance in use of these features has not been assessed, this basic approach appears to be sound. The main characteristics of these units are illustrated in Figure 4.23.

Flush copings. Although there is clear evidence that overhanging copings help minimize staining of a building façade, many designers still prefer a choice so they can employ flush details to fit in with design strategies that require plain and featureless elevations. Although these flush features will always be more prone to disfiguring staining than those that overhang with throating beneath, there are a number of techniques that can minimize these problems:

- Provide a steep fall-away from the elevation that is to be the most prominent. This could be achieved through a specially manufactured unit that allows for a step down (Figure 4.24).
- Ensure that the amount of water settling on the surface of a coping flows away from the prominent elevation by fluting or grooving the surface. This was first observed on a building in York (Figure 4.25).

Here the flutes were taken down to discharge over an observable face of the coping which detracted from what could have been a more successful method of directing water flow. This has been refined to produce a more effective fluted coping, illustrated in Figure 4.26.

4.5.3 Run-off at joints

These surface improvements still do nothing about the problem of intensified flow down and through the joints of the coping. Traditional construction books refer to 'joggle joints' between stone coping units. Figure 4.27 is an extract of one such detail. Essentially this was an overlapping joint that would restrict direct through-joint drainage. Further development of this feature using some of the earlier observations has produced a more

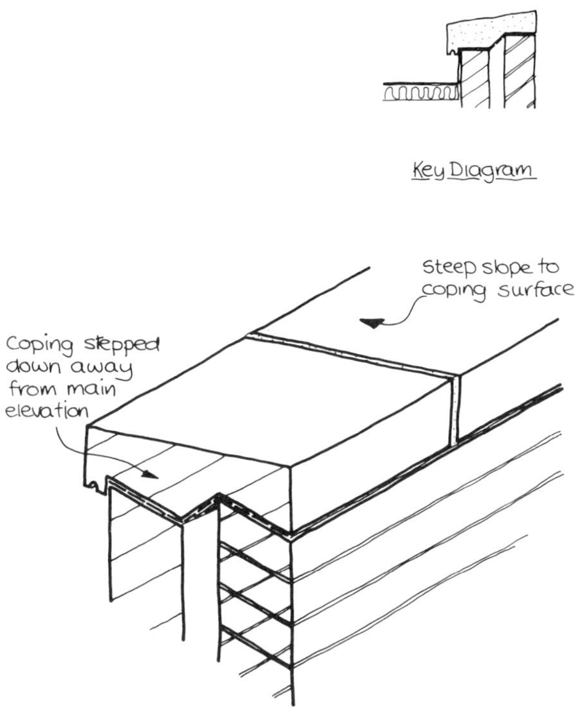

Figure 4.24 Possible design of a stepped steeply sloping coping.

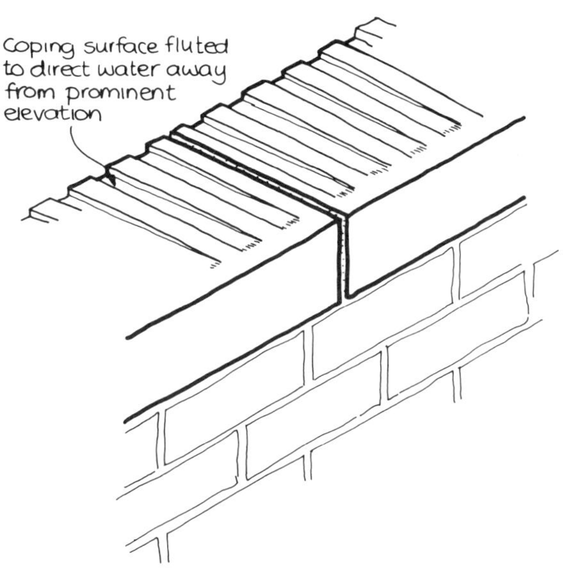

Figure 4.26 Possible design for a flush coping with a sloping fluted top surface.

Figure 4.25 This wide inclined stone coping has its top surface fluted which helps to keep water flow down the prominent elevation to a minimum. However the flutes were still allowed to discharge over an observable face causing ugly streaking.

sophisticated joggle joint that could effectively resolve joint staining problems. This is illustrated in Figures 4.28 and 4.29 and has a number of unique features:

- the joggle joint extends the full width of the joint;
- the joint surface is inclined towards the less prominent elevation;
- the 'underlapped' surface is grooved or recessed to minimize the movement of water within the joint.

This type of joint has a number of implications:

- The designer has to accept the visual impact of an overlapping joint on the face of the coping although this could be restricted by stopping the joint short by 50 mm.
- Most of the increased joint flow will be directed to the less prominent face of the coping and although the throating underneath the rear overhang will prevent much of the transfer to the wall surface, an increased risk of staining must be accepted.

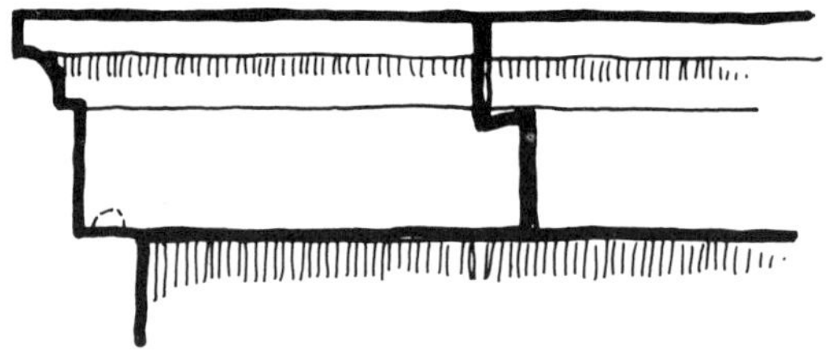

Figure 4.27 Sketch showing a traditional joggle joint to a stone coping.

Manufacture of such a coping may be problematic as tolerances would have to be much finer and the overlapping components may be vulnerable to accidental damage during delivery and installation. Dry jointing is a possibility, minimizing throating blockage and cutting down on lime staining.

Where a designer wants a flush face to the coping, water run-off can be further controlled by the top surface being fluted as previously described (Figure 4.30).

Although these designs are untried and untested, the key features point towards copings that could help to avoid some of the many past failures.

4.6 ROOF EAVES AND VERGES

4.6.1 Verges of industrial and commercial buildings

The verge detail of many roofs to commercial and industrial buildings is usually protected by pre-formed verge sections that are coordinated in their design with the main profiled metal roof sheeting. They tend to extend across the roof slope by some 100–150 mm and down the wall elevation by a similar amount. This verge component has no drainage function, it merely protects the junction of the roof and wall elements. Because of the width of the horizontal component, a considerable amount of water can flow down the slope. Where the different sections overlap or butt up to each other, capillary action will draw water flow down on to the face of the wall. Figures 4.31 and 4.32 show an example typical of this effect. This form of staining is surprisingly common on many industrial and commercial estates.

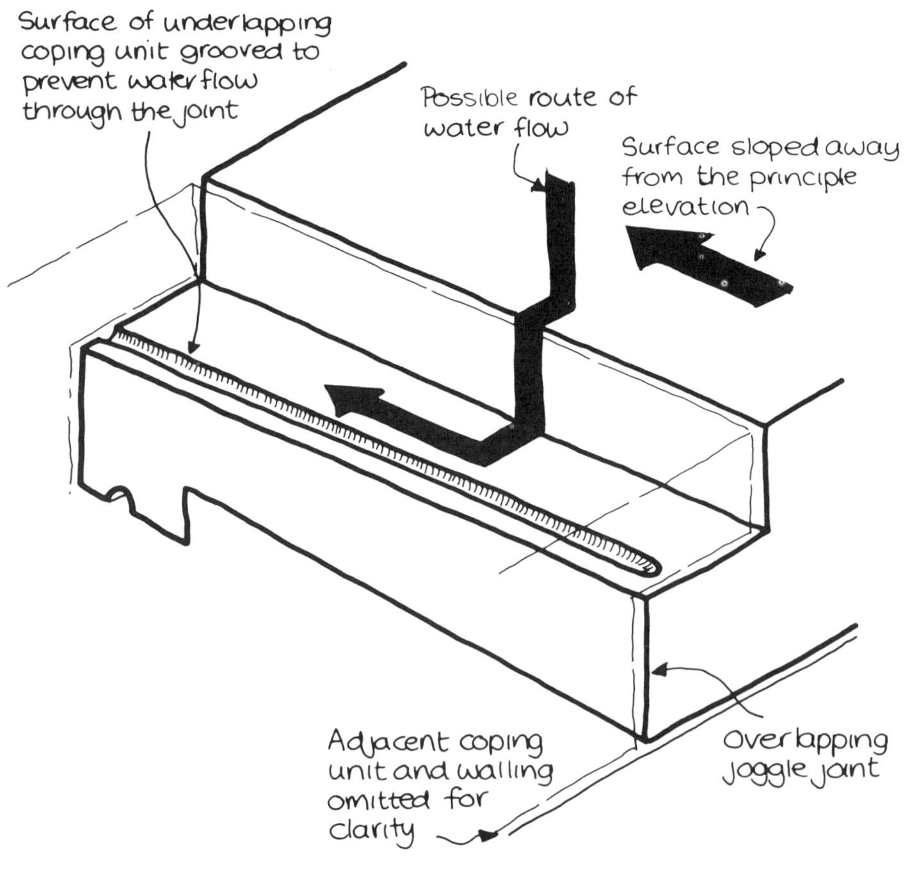

Figure 4.28 Diagrammatic illustration of a proposed joggle-jointed sloping coping.

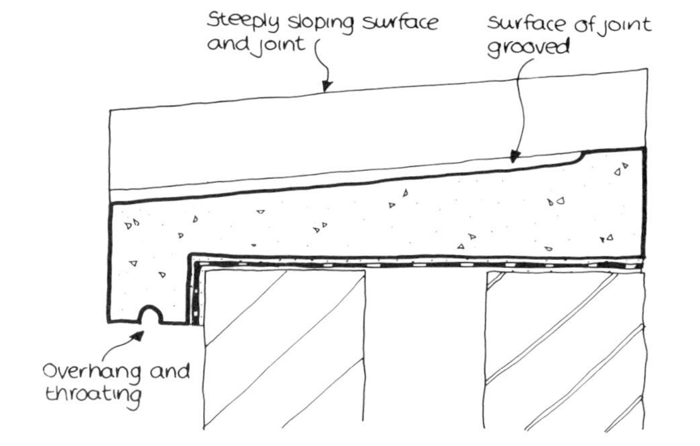

Figure 4.29 Sketch section through joggle-jointed sloping coping.

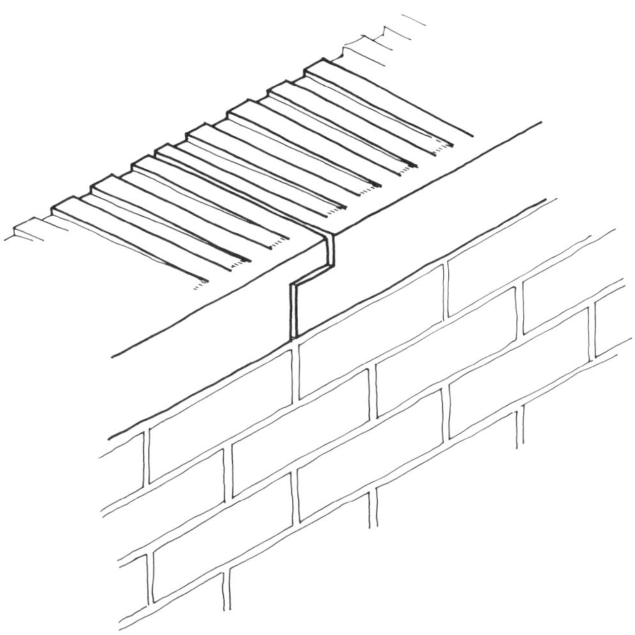

Figure 4.30 Illustration of a flush joggle-jointed fluted sloping coping.

Figure 4.31 Staining at the joints of pre-formed metal verge cloaking sections.

Figure 4.32 Detail of Figure 4.31 showing high levels of water flow at junction of adjacent units.

4.7 PLAIN FAÇADES

Staining is often associated with features on a façade that interrupt the flow of water or introduce it in a more concentrated form. But plain façades devoid of any architectural features or detail often pose an equal problem. A wall will never weather evenly and a large plain façade constructed of the same material will act as a canvas on which the environmental agents will paint a sometimes unflattering image. These façades have no relief, detailing, horizontal components or changes of material that will help to carry any staining. The type of effect will depend on the material used; Verhoff (1988, p. 142) has categorized these according to absorbency, as follows.

4.7.1 High absorbency façades

Because these façades will rarely receive enough rain to allow water flow to develop, the material close to the top of the façade will receive the most water and so change the most in appearance. In these locations it is vital that effective copings are provided to minimize the amount of water flowing off the parapet and/or roof. This coping must direct the rainwater away from the exposed façade and towards the roof. On more sheltered façades, although the amount of water will be less, the façade can remain damp for much longer as there will be less sun and wind allowing algae to develop.

4.7.2 Medium to low absorbency façades

Here water flow will develop in the right circumstances but not extend all the way down to the ground. In other words, the façade will not be completely washed as the flow will cease where the rain hitting the surface is insufficient to maintain flow. It is at this point that dirt carried down from higher levels will be deposited. The only way to control this type of staining is to limit the flow. Horizontal string courses bold enough to throw water off the façade are proposed by Verhoff as they can serve several functions:

- reduce the amount of water on the surface;
- reduce the differences between panels at different levels on the façade;

- make the change from washed to unwashed gradually by allowing architectural detailing to produce visual distractions and shadow areas that can mask staining.

In many ways the Perret buildings in Le Havre use this approach (see Figures 2.1 and 2.2 in Chapter 2).

4.7.3 Non-absorbent façades

Rain hitting non-absorbent glass or curtain walling will often flow right down to the bottom of the building. If this happened in an even film then it would be self-cleansing and give rise to few aesthetic problems but because water will form into discrete streams, these façades can become streaked, especially in heavily polluted areas. Whereas the more absorbent materials can carry a certain amount of dirt, these non-absorbent façades are often designed to retain a clean and bright image, therefore even a small amount of dirt staining may detract from its intended visual impact. In many locations these façades may need to be cleaned at regular intervals.

4.7.4 Good practice

Because of the range of visual effects that may occur on large areas of plain walling it is unwise to include these features in a building's design. Higher-value claddings such as stainless steel may be able to retain a consistent visual image across a greater area but any porous material will be adversely affected. Consideration should be given to breaking up the visual impact of an elevation by varying material colour and textural quality, horizontal string courses and other features, glazing, set-backs, etc. But care must be taken that any feature that is incorporated does not produce its own staining problems.

4.8 WINDOWS AND GLAZED PANELS

4.8.1 General aspects

Windows, glazed areas and other non-porous panels pose the designer with special problems because:

- compared to the adjacent walling surface they hold little water other than the amount of droplets on the surface;
- rainfall run-off quickly occurs during even the weakest of showers;
- non-absorbent materials are often expected to be washed clean by the rain yet most are hydrophobic, resulting in water running in discrete streams or rivulets which can cause a streaked effect;
- the high levels of run-off that flow away from non-absorbent areas, if allowed to run over the walling below, will either wash the surface clean or result in staining. Which one depends on a variety of factors including exposure, material characteristics, building detail, etc. In most locations in this country, walling below windows and glazed panels that are facing south and west will be washed clean while those on the north and east elevations tend to be affected by dirt deposits and algae growth.

4.8.2 Windows and cills

The presence and effectiveness of any cills beneath windows and other glazed features in an elevation will determine whether the walling is washed or stained. Addleson and Rice (1994, p. 573) pointed out that effective cill projections can allow dirt to accumulate on the wall surface below. Where this matches the shadow area cast by the sun then it will not be so noticeable but where it extends further than this, it can disfigure the elevation. Where cill projections are ineffective or absent then the high levels of run-off will variably affect the walling below, producing areas of variably stained and washed wall.

The extent to which this will happen will depend on a number of factors:

- the total amount of rainfall on the elevation;
- the absorption characteristics of the walling below the glass;
- the effectiveness of the cill (if any!).

This is one of the most difficult judgements to make when detailing a building.

4.8.3 Cills without stooling

It has long been recommended practice that cills below windows should be 'stooled', or have their surfaces recessed, leaving upstands at either end to ensure that water does not flow off the edge of the cill during wind gusts, bypass the throating and affect the wall surface (Figures 4.33 and 4.34). Yet many designers still include cills without this feature that will give rise to a 'moustache' at either end of the cill. The building featured in Figures

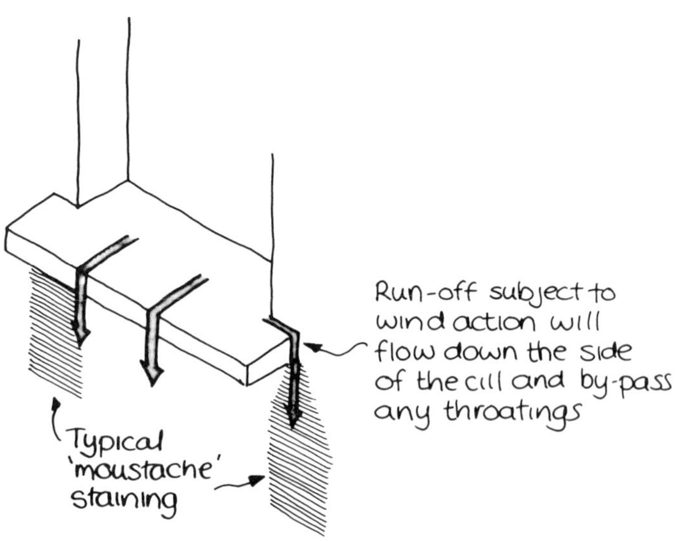

Figure 4.33 Typical 'moustache' staining of a cill without stooling.

4.35 and 4.36 shows such a staining pattern. This visual effect is probably enhanced by the detrimental effects of the sealant around the window frames. Stooling is equally important for metal cills where the feature takes the form of a metal flange.

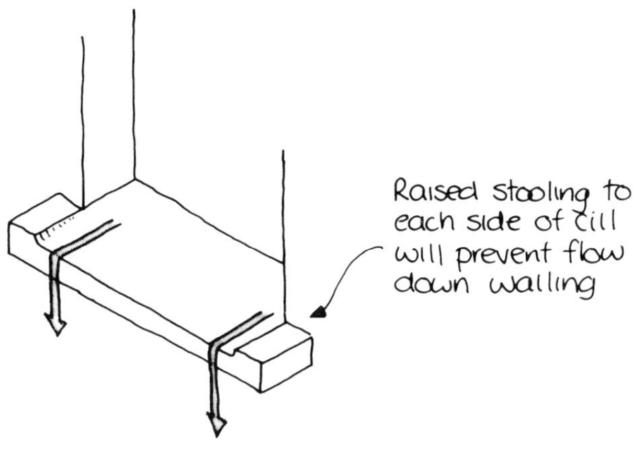

Figure 4.34 Cill with stooling.

Figure 4.35 Recently refurbished building showing 'moustache' staining.

Figure 4.36 Detail of Figure 4.35. This staining has probably been enhanced by the breakdown of the sealant around the frame depositing substances on the walling.

4.8.4 Large glazed areas and curtain walling

Many new buildings, especially those in commercial districts, have made significant use of curtain walling to the extent that complete elevations and sometimes whole buildings are clad in glass. In relation to staining, previous discussions have revealed that glass presents its own problems but where elevations incorporate a mixture of façade materials, more prominent staining problems can occur. Typical examples would include an area of curtain walling discharging uncontrollaby over stone or brick cladding at a lower level. In these positions higher levels of water run-off than would otherwise occur will flow over the more absorbent cladding beneath. This will result in variably stained walls. The new commercial building in Figures 4.37 to 4.39 shows how this occurs. The fully glazed corners of the building allow a considerable flow of water to build up. This accumulates and passes over the brick cills at the fourth and second storeys. Because these cills do not have the ability to shed water, all the increased flow is discharged over the brick. Although the building was only recently completed, Figure 4.39 shows the beginnings of efflorescence along the edges of the damp areas, which will soon be followed by the other agents that cause premature staining.

Figure 4.37 General view of a recently completed commercial building with fully glazed 'corners'.

Figure 4.38 Detail of the bottom two storeys showing the lack of any cills at the glass/brick cladding junction.

Figure 4.39 Early signs of dampness and efflorescence to the brick cladding at the base of the building.

Where the curtain walling mullions are pronounced externally, the actions of the wind will drive flowing water down across the surface of the glass into concentrated streams that follow the line of these mullions. If the cill at the base of the glazing is not effective at shedding water, then soiling will be concentrated in regular streams on the walling below (Figure 4.40). The buildings in Figures 4.41 to 4.43 illustrate this effect very clearly.

4.8.5 Good practice

In both these cases there can be no substitute for a well-designed stooled cill that prevents this excess of water from affecting the cladding surface. In addition, consideration should be given to:

- extending the vertical component of the glazing down to the ground level so water does not have to flow over another form of walling;
- below the glazed cladding areas, using materials of a texture/colour that are able to mask or make the resultant staining less prominent.

4.8.6 The effects of sealants

Many cladding systems use sealing compounds and gaskets to ensure water tightness and, on some buildings, dark staining has been observed directly below the position of these sealants. This effect has been investigated by Nakayama and Takei (1988): from tests on a number of natural stone panels jointed with polysulphide, modified silicones and silicone sealing compounds, it was found that the cause of the stains was the migration of silicone which bled from the sealing compound onto the adjacent stone surface near the joint. Once deposited, dirt and soot adhered to the silicone, making the stone surface more unsightly.

Improvement in the performance of sealants is in the hands of the manufacturers who must ensure that their products are stable and will not allow migration to affect adjacent surfaces. Observations have shown that fully exposed joints, where the sealants and gaskets are open to the elements, appear to be more affected than those that are physically concealed. Figure 4.44 shows sealant staining on the surface of darkly coloured polished stone cladding that would have been more prominent had the cladding been lighter in colour. The moustache staining on the building in Figures 4.35 and 4.36 has probably been caused by sealant migration.

4.9 SIGNS, LETTERING AND OTHER SMALL PROJECTIONS

4.9.1 Signs and lettering

Many building owners who are keen to make a feature of their corporate identity often put their logo or name on

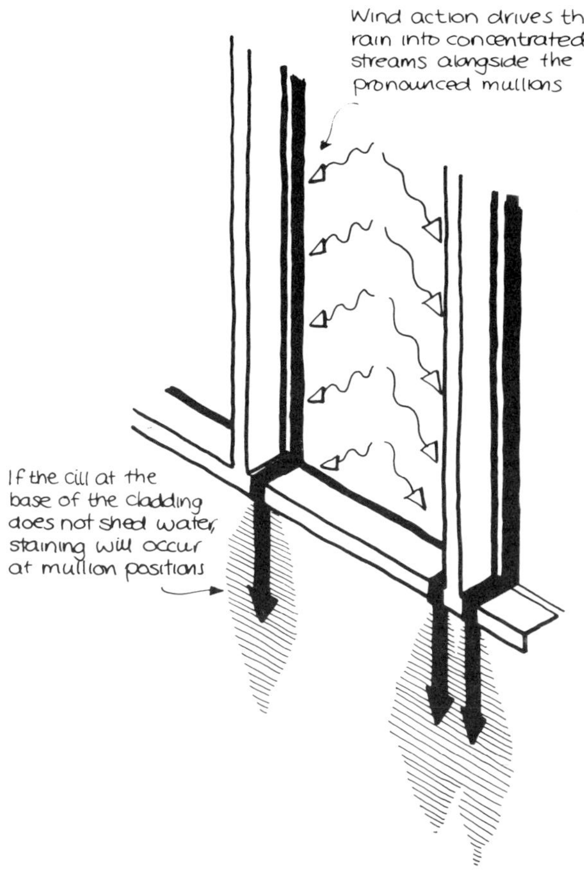

Figure 4.40 Diagrammatic sketch of water flow on glass cladding.

Figure 4.41 General view of multi-storey educational building clad with curtain walling with pronounced mullions.

Figure 4.42 Detail at junction of curtain walling and concrete plinth showing concentrated staining at mullion positions.

the front façade of their building. It is often prominent, designed to be seen by users and the public – in short, intended to make a visual impact. The main problem with this is that any projection from a façade will interrupt the flow of water and redistribute it, washing in some places and depositing dirt in others, causing an unsightly effect. The visual prominence of signs and lettering makes this form of staining even more noticeable.

Observations of logos and lettering that are fixed back flush to the façade have shown that they will cause staining through a number of mechanisms:

- The flat and slightly sloping surfaces of the objects will accumulate higher levels of dirt and other pollutants that will be washed off and deposited on the façade at a lower level.
- Rainfall hitting the exposed horizontal surfaces can bounce back on to the façade and wash the wall above clean in small patches.
- Certain letters can concentrate run-off by the nature of their shape. The letters 'V', 'W', 'X', 'Y' and 'U' are particularly susceptible. This is often enhanced where the junction to the wall is not properly sealed, so that water can run behind the sign and wash or stain the wall at a lower level where the façade would otherwise have been sheltered.

Figure 4.43 Staining at mullion positions where cladding is less pronounced.

Figure 4.44 Staining caused by sealant or gasket migration.

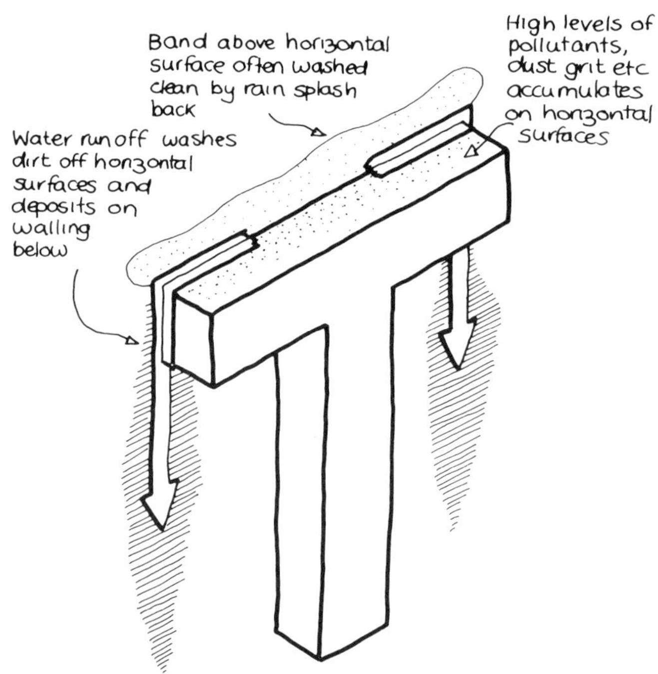

Figure 4.45 Diagram showing the variety of staining and washing effects associated with protruding signs and lettering.

Figure 4.45 illustrates this effect and the buildings shown in Figures 4.46 to 4.51 shows the visual impact that can result.

Fixing the sign or lettering away from the façade surface can help to minimize the variable washing. For example, the logo and lettering shown in Figures 4.49 and 4.50 demonstrate the difference between the lettering that is fixed flush and the logo that is fixed proud of the wall surface, and has yet to be affected.

4.9.2 String courses, cornices and other ornamental bands

These largely ornamental features were an integral part of building design up until the early part of this century (see Chapter 2) when unnecessary decoration became less fashionable. In this post-modern age many designers

Figure 4.46 General view of a new hotel (three years old).

Figure 4.47 Detail of lettering fixed flush to the wall surface. Evidence of early staining especially below the 'V'.

Figure 4.48 Logo and lettering on the elevation of a five-year-old hypermarket.

Figure 4.49 Detail of Figure 4.48 showing staining below lettering that is fixed flush to the wall.

Figure 4.50 Detail of logo showing the different fixing methods used for the logo and lettering. The logo has been fixed approximately 75 mm away from the face of the wall and is yet to be affected by staining. The lettering is fixed flush and already shows signs of disfigurement.

have felt able to incorporate these horizontal features in their designs once again. They can be useful in breaking up the visual impact of an elevation and allow for a change in façade material, colour or texture. In section 4.7.4, string courses were also identified as features that can help mask the effects of the staining process.

Despite these positive advantages it is important to ensure that these features are designed correctly; if not they can be the cause of staining problems rather than the solution. The building in Figures 4.52 to 4.56 used light coloured string courses and cornices as central features of the design. Of these, the strongest aesthetic impact is the upper cornice positioned just below the roof eaves (Figure 4.53) and the design of this creates a number of problems:

- It is very light in colour making it very vulnerable to disfigurement.
- Although it is weathered it provides a relatively broad upper surface on which high levels of pollutants can collect. It also provides an excellent roosting place for pigeons who add to the debris in their own particular fashion!
- The water run-off is allowed to flow over the edge of the cornice which does not have the benefit of a throating underneath. The water flow continues over

Figure 4.51 Logo and lettering fixed flush to stone cladding causing variable deposition and washing.

Figure 4.52 General view of building with cornice and string courses.

Figure 4.53 Detail of strong visual cornice units. Note that the side facing the camera has been netted to prevent pigeon roosting while the other has not.

the whole feature depositing dirt and greatly disfiguring it.

- Because of enhanced flow of water at the joints the parts of the coping units that are directly below are washed clean while the adjacent areas remain dirty, which contributes to its unsightliness. This joint flow is continued down the face of the brick walling and concentrated further by the influence of the perpens of every alternate brick course until it reaches the lower, less prominent string courses. Here, because the joints in these bands are coordinated with the joints in the cornice, this joint flow is confirmed. This has resulted in long vertical stains down the whole face of the building, a common occurrence on many building types where joints in components are in the same vertical plane.
- On one elevation for some unknown reason, two of the cornice units have been provided with throatings

under the leading edge (Figures 4.55 and 4.56). The visual effect is dramatic. The two units with the throating have remained relatively unblemished while those to each side have been heavily stained. This shows the importance of the throating. The diagram in Figure 4.57 summarizes this process.

String courses on other observed buildings have been slightly more successful. In a number of cases these ornamental features have been made weatherproof by the addition of a lead cover flashing over the whole of the upper surface. Although this still allows the water to flow over the edge, if positioned and detailed correctly it could prevent the problematic through-joint flow. The building in Figure 4.58 seems to be a promising example. Here the lead has been dressed over the leading edge of the string course and welted at regular intervals but not at the same spacings as the joints in the string course itself. This last feature is important as this will help to prevent through-joint flow. A speculative drawing of this detail is shown in Figure 4.59. Other positive features of this detail include:

Figure 4.54 General view of the cornice and string courses showing the beginnings of staining concentration following the joints.

Figure 4.55 General view of the elevation containing two cornice units that have had a throating provided under the leading edge.

Figure 4.56 Detail of Figure 4.55 showing the clean surface of the two cornice units with throatings compared to those without.

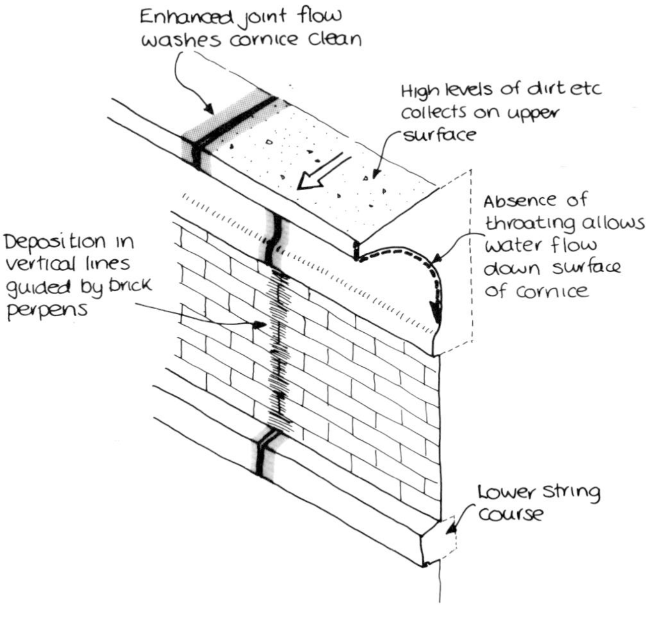

Figure 4.57 Diagrammatic sketch of the staining mechanisms illustrated in Figures 4.52–4.56.

Figure 4.58 Part elevation of a new city centre building with lead-covered string course.

Figure 4.60 Detail of string course showing wiring to prevent pigeon roosting.

Figure 4.61 Part elevation showing discontinuity of string course above entrance area.

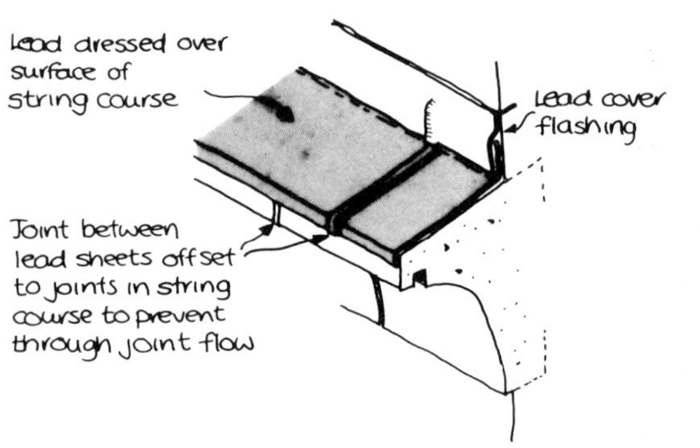

Figure 4.59 Speculative sketch of string course in Figure 4.58.

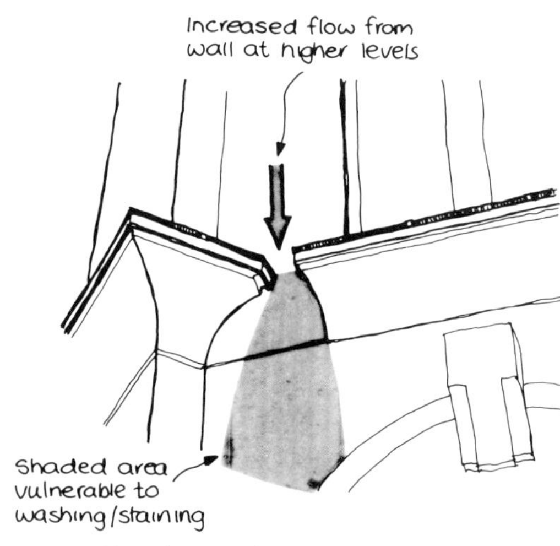

Figure 4.62 Sketch showing area of walling at risk of disfigurement in Figure 4.61.

- a high degree of accuracy of manufacture and construction which helps to maintain throating integrity;
- the provision of proprietary wiring above the top surface to prevent pigeon roosting (Figure 4.60).

The one feature that lets this elevation down is the way the string course is detailed above the entrance door. By providing a physical gap in the course, water flow from higher levels will extend down and variably stain or wash the walling below the string course. Although this building is very new, streaking has already occurred in this area. Figures 4.61 and 4.62 show this in detail.

4.10 EXTERNAL ENTRANCE AREAS

The entrance of a building is the focal point of the principal façade and is the first visual impression that most users and visitors experience. Where a building has a commercial function then this can be a very important feature in the design.

On many of the buildings observed as part of this study, entrance areas were often poorly designed and detailed as if they had slipped the mind of the designer. As a result, ugly and prominent staining can be the first feature that a person visiting the building will notice. The office complex illustrated in Figure 4.63 is a typical example and shows how this could affect the commercial viability of a building. When there is a surplus of office buildings in an area, who is going to buy or lease one that has turned a slimy shade of green within the first six months? Figures 4.64 to 4.72 illustrate some of the adverse features in more detail.

Appraising this and other buildings, a number of features that cause recurrent problems can be summarized:

- steps and staircases
- parapet walls
- canopies and small roofs
- hardstanding areas
- planted areas.

These will be considered in more detail.

4.10.1 Steps and stairs

These features, especially when fully exposed, will always pose the designer with a problem. In addition to rainwater run-off and normal atmospheric pollutants, steps and stairs will be soiled by the debris and dirt deposited on them by the feet of the building's users. This will be washed down the stairs and deposited on the risers and the treads. The risers are normally the worst affected, with large amounts of dirt collecting at the joints or masonry perpens (Figures 4.73 to 4.76). Actions to minimize this could include:

- provision of a canopy over the stair to reduce the amount of water affecting the surfaces;

Figure 4.63 General view of a speculative office building adjacent to a busy motorway junction. Because of the difference in levels from the street and the entrance there is a considerable amount of staircases, ramps, planted areas, canopies. etc.

Figure 4.64 Much of the hardstanding is drained by discharging over the face of the retaining walls.

Figure 4.65 Detail of Figure 4.64 showing high levels of disfigurement.

Figure 4.66 Light-coloured coping units have been used around all the brick-built planters.

Figure 4.67 The coping units have rounded edges with no throating beneath allowing water to flow down the surface of the walling below.

Figure 4.68 Conspicuous joint staining has affected all parapet walls.

Figure 4.69 In many areas poor damp proofing behind retaining walls has allowed intense lime staining to disfigure the brickwork.

Figure 4.70 An isolated brick plinth has been badly stained by water flow from the hardstanding above and spitter pipes draining the adjacent planters.

Figure 4.71 The watertightness of virtually all spitter pipes was defective allowing ugly water flow down the face of the brickwork. Here the water is draining around the pipe rather than through it.

Figure 4.72 The canopies above the main entrances have no guttering which allows rain splashback to stain the walls at low level.

- a drainage channel at the top of the staircase to ensure minimal flow down it;
- overhanging and throated treads that minimize the direct flow of water down the face of the riser (Figure 4.77). This design has not been manufactured or tested but could indicate the usefulness of trying to minimize water flow directly down the face of the risers;
- a colour choice that will make the soiling less noticeable;
- regular cleaning of the stair.

The junction of the staircase and side walls can also be badly affected by splashback during rainstorms and will stain the wall as well as promote frost damage. A skirting or upstand of impervious material is essential here (Figures 4.78 and 4.79).

4.10.2 Parapet walls

The rules regarding free-standing parapet walls to staircases are the same as for parapet walls in any other location but with a few special features:

Figure 4.73 External entrance stairs to an arts centre consisting of brick treads and risers.

Figure 4.74 Detail of Figure 4.73 showing accumulation of dirt at joints especially to risers.

Figure 4.75 Entrance stairs to a leisure centre constructed from pre-formed dressed stone risers and treads.

- *The handrail or baluster* – many designers use a dwarf parapet wall with vertical balusters/posts extending the handrail to the appropriate height and are fixed either to or through the coping feature. This will provide additional points of entry for water to bypass what otherwise would be a waterproof barrier. The handrails should be fixed to the surface of the walling below the coping system if possible (Figure 4.80).
- *Change in inclination* – at the base of the staircase the parapet wall will terminate and unless the high levels of water flow down the coping are properly dealt with then the concentrated run-off can cause ugly staining (Figures 4.81 to 4.83).

It is important to have well-designed copings that shed as much water as possible with the special attention to the coping unit that spans across the change in the inclination. Thought must be given to which elevation of the staircase is the most prominent, and the coping run-off directed to the other face (Figure 4.84).

4.10.3 Canopies and small roofs

Many entrance areas utilize small canopies and roofs either for aesthetic effect or to provide shelter for visitors and users of the building while they gain access to the building. Many appear to be afterthoughts without

Figure 4.76 Detail of Figure 4.75 showing that despite the overhanging nature of the treads, the risers are still badly soiled.

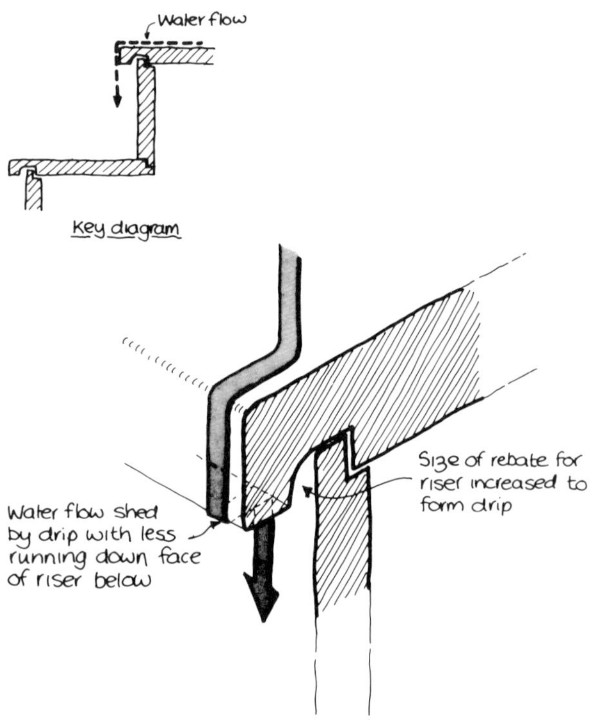

Figure 4.77 Diagram showing possible location of throating to underside of tread to prevent water flow down the face of the riser. *Note*: weakness to the edge of the tread would have to be resolved!

formal drainage systems, which will inevitably result in the uncontrolled flow of water over adjacent surfaces. The features illustrated in Figures 4.72, 4.85 and 4.86 are clear examples of this practice.

Where any canopy or roof surface is provided, no

Figure 4.78 Impervious tiled treads and risers will allow considerable splashback to stain adjacent brickwork. The light colour of the tiles will show dirt clearly and the smooth surface makes them very slippery when wet!

Figure 4.79 Longer exposure to the elements will lead to algae and moss growth at the junction of the steps and brick wall.

matter how small, it should always be drained directly to the surface water drainage systems in a controlled fashion. This means that decisions on these features must be taken early on in the design process so that drainage provision can be made.

Figure 4.80 Staircase where the handrail and balusters have been fixed through the coping construction. The junctions of the main posts and the copings are very difficult to waterproof, potentially allowing water to stain the walling below.

Figure 4.81 The broad flat surface of the parapet wall to this staircase (see Figure 4.80) allows large amounts of water to flow down to the base of the stair and discharge over the walling.

Figure 4.82 Brick on edge coping and tile creasing fails to prevent the increased flow of water at the change in inclination. Lime staining is also evident here.

4.10.4 Hardstanding

Similar lack of attention to drainage of paved hard-standing areas around entrance areas has also been noted. In some cases water has been directed to flow over the face of prominent elevations to adjacent retaining walls (Figures 4.64, 4.65 and 4.85). A more appropriate approach would be:

- Ensure all hardstanding areas are formally drained via a yard gulley or similar.
- All gradients on the hardstanding should fall away from prominent elevations.
- Where an area cannot be drained in a proper manner, a projecting feature should be provided to shed the water away from the walling below in a less prominent position.

4.10.5 Planters

Plants, bushes and trees have regularly been used to make entrances to buildings more attractive. This creates a number of particular problems:

- The presence of vegetation will result in a greater range of organic growth occurring on building materials.
- Depending how much the vegetation grows over adjacent surfaces, variations in shelter offered by the plants/trees will affect how the building fabric changes. Shaded areas will remain damp for longer periods but will be washed less by rain run-off.
- All planted areas need to be well drained but unless

Figure 4.83 The 'clean lines' of this design has resulted in the absence of any form of overhang/throating allowing water flow to affect the face of a brick-built bin store below.

Figure 4.85 This glass-covered canopy has no guttering system and discharges over the flat entrance area below. Because there is no raised upstand on this balcony the water flows over the walling below causing staining.

Figure 4.84 The broad flat coping at high level allows water to flow down the face of the prominent elevation disfiguring the reconstituted stone blocks.

Figure 4.86 Similar effect to Figure 4.85 but on a smaller scale.

special care is taken to ensure that the containers are properly waterproofed the visible surfaces of the enclosing structure will be disfigured (Figures 4.66 and 4.70).

In many schemes the waterproofing specification for the internal surfaces of planters has been minimal, involving the application of one or two coats of bitumen-based paint. The aggressive environment of planted areas will soon render this ineffective, increasing the risk of serious staining. The standard of waterproofing should match that for any flat roof or usable basement area, with particular care taken to use materials that can withstand an organic environment.

REFERENCES

Addleson, L. and Rice, C. (1994) *Performance of Materials of Buildings*, Butterworth–Heinemann, Oxford.

Andrews, A. (1994) *Stone Cleaning. A Guide for Practitioners*, Historic Scotland, Edinburgh.

BSI (1983) *BS 5642: Sills and Coping. Part 2: Specification for Copings of Precast Concrete, Cont Stone, Clayware, Slate and Natural Stone*, British Standards Institution, Milton Keynes.

Dunster, D. (1988) *Architects Journal*, **188**(39).

Hawes, F. (1986) *The Weathering of Concrete Buildings*, Cement and Concrete Association, Slough, p. 2.

Nakayama, M. and Takei, Y. (1988) *Stains Caused by Sealing Compound on Stone Cladding*, Kajima Institute of Construction Technology, Tokyo.

Verhoff, L.G.W. (ed.) (1988) *Soiling and Cleaning of Building Façades*, Report of the Technical Committee 62 SCF, RILEM, Chapman & Hall, London, p. 142.

Chapter 5
Establishing a design strategy

5.1 INTRODUCTION

The previous chapters attempted to:

- establish premature staining of new buildings as a performance problem that needs to be addressed by the design profession;
- review some of the mechanisms of staining;
- through observations, identify some of the most common staining problems and propose building details and methods that may limit the extent of the disfigurement.

This final chapter will attempt to outline a series of strategies or models that may help the designer formulate solutions to minimize the possibility of their buildings falling victim to premature staining.

5.2 DESIGN STRATEGIES

When considering how the visual impact of a building can be protected and enhanced over time, a designer must consider more than just what façade material to choose or how to detail the elevations. A broad range of factors determine whether a building will stain or not; these must all be accounted for at the appropriate stages in the design process, including inception (client's briefing, what sort of building is required, life span of building, costs, etc.), the feasibility stage (site influences, prevailing winds, climate, etc.), through to the detailed design phase (design of cills and copings, material choices, etc.).

Before a checklist of factors is outlined, it will be useful to consider a few conceptual models of how buildings perform over time to put the issue in context. From the research reviewed in this study, three approaches have been developed:

- Verhoff (1988) looked at the visual quality of a building over time.
- Andrews (1994) linked soiling with aesthetic value and building form.
- Atkinson (1977) looked at the components of the performance and appearance of wall surfaces in systems terms.

These approaches are not mutually exclusive; there is considerable overlap and not one single model provides all the answers, but they help designers visualize the problem and conceptualize what level of performance they want from their proposed buildings over time. Once this is clear, then resolving the component parts of the problem can be a more manageable task. If designers do not know where they are going with a design, how do they know when they have arrived?

5.3 VISUAL QUALITY OVER TIME

5.3.1 Introduction

Verhoff (1988, p. 58) suggested that there are three approaches:

- *Strategy 1* – design buildings that have eternal youth.
- *Strategy 2* – produce buildings that can be brought back to their original appearance at regular intervals by the injections of additional finance.
- *Strategy 3* – design buildings that grow old gracefully without expensive maintenance.

These strategies are illustrated in graphical form in Figure 5.1.

5.3.2 Defining terms

Before each strategy can be discussed in detail, the terminology needs to be defined.

Visual quality. 'Visual quality' is a subjective concept but for the purposes of this study it could be defined as the visual impression at the time of completion that was envisaged at the design stage. As few buildings ever achieve this ideal state because of efflorescence, poor workmanship, etc., the finished product will not start at the ideal level. Soiling and staining will generally reduce visual quality levels.

Threshold of acceptability. Acceptance threshold is another subjective value and its definition is a topic that

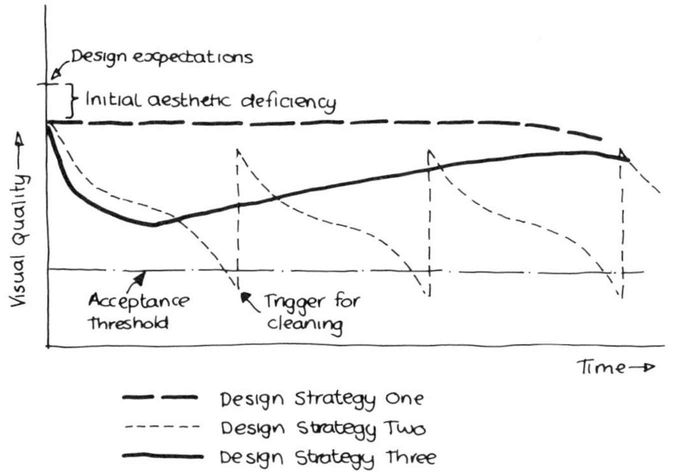

Figure 5.1 Graphical representation of visual quality of a building over time (Verhoff, 1988).

has received little attention in the research community. Here it is defined as the level of soiling and disfigurement that is considered as just aesthetically acceptable by a broad community of designers, owners, users and the general public. Below this level, a decision to clean or alter the façade to remove or prevent the staining would be triggered.

Time scale. The life cycle for most buildings has been traditionally set at 60 years after which the whole building would be so structurally and functionally redundant that demolition would be the only option. 'Normally' is the key phrase here; technological changes and an active property market during the 1980s rendered many buildings functionally redundant resulting in their demolition in as little as 15–20 years after they were built. Therefore the time scale on this model must be related to the actual building; those in urban/commercial centres that command high rates of return may well be quickly outdated as markets develop and change, whereas buildings in more stable economic environments may enjoy greater longevity. Such a difference in the expected life of a building may well affect design decisions in relation to staining.

5.3.3 Design strategy 1

The bold broken line in Figure 5.1 represents a building that is designed for eternal youth, resisting the passage of time and the attempts of the weather to alter the appearance of the building. This could only be achieved by:

- expensive and durable façade materials;
- careful and detailed design of the façades;
- thorough and regular cleaning and maintenance programmes (cleaning in this case would be a routine operation to remove initial deposits of dirt, similar to window cleaning operations).

The cost of this approach would be very high and therefore limited to the most prestigious of buildings where the commercial return (or public subsidy) is likely to justify the high initial expense. Examples include the Lloyds building (Figure 5.2) in the City of London where stainless steel cladding has been used to produce a building in which parts of the elevation weather very well. But this level of performance is not matched by all the components of the building; Figure 5.3 shows typical staining to part of the exposed concrete frame. Verhoff (1988, p. 59) also asked the broader question of whether society wants buildings that stand apart from their environment, never becoming part of the mature and familiar backcloth 'which has been bequeathed to us by past generations'.

5.3.4 Design strategy 2

The lighter broken line illustrates the second approach, by which buildings are brought back to their original appearance at regular intervals by the injection of additional money for cleaning, repainting or both. A traditional example of this would be the rendered Georgian building in many town centres that is repainted to enhance its attractiveness; it is also a useful way to revitalize whole areas. But there is a conceptual problem with Verhoff's original strategy because although the

Figure 5.3 Detail of a junction to the exposed concrete frame of the Lloyds building showing uneven washing and staining.

Figure 5.4 General view of stained law court building.

building's appearance will 'peak' after cleaning or painting, it will never reach the original starting point. This is because no matter how well a building is cleaned or repainted, it will never achieve its original aesthetic qualities. The Masonry Research Group in Scotland (Andrews, 1994) has recorded numerous examples of sandstone buildings that have been badly affected by the cleaning process where the resulting surface was permanently disfigured by previous water run-off, general deterioration or the cleaning process itself. A contemporary example is shown in Figures 5.4 to 5.8 which stained very badly within a few years of its construction. The building was cleaned but a clear impression of the former staining still remained (Figures 5.9 to 5.11). If more rigorous cleaning methods had been used, the stains might have been removed but the façade material could have been irreversibly damaged. Therefore the graphical profile for this strategy should include a diminishing visual quality over time even after cleaning (Figure 5.12). Depending on the circumstances, after a certain length of time the benefits of rejuvenating the façades would be so marginal that many building owners would be reluctant to carry it out at all. Another drawback about this strategy is that the building would look 'down at heel' and in need of maintenance for a significant proportion of its 'cycle'. Typical examples of buildings that might fit into this category include those with façades of low to medium quality materials, little care taken during detail design stage, plain elevations with few features, building owners with short- to medium-term outlooks and little or no preventive maintenance or cleaning programmes.

Figure 5.5 The glazed roof and walls to the main entrance and staircase has allowed high levels of water to flow over the stone cladding beneath. The cladding was soon disfigured by algae growths.

Figure 5.6 The inclined cills beneath the glass cladding concentrated the water flow at the change in inclination.

5.3.5 Design strategy 3

This is represented by the full line on Figure 5.1 and is an attempt to design buildings that grow old gracefully without expensive maintenance. There is an acceptance that the façade of the building will change but in a way that will enhance rather than spoil it. In the initial stages, early staining may have a detrimental effect as any soiling of a new façade will be prominent but as the building ages this becomes more evenly distributed and so less visually offensive. In Verhoff's opinion, this strategy is probably the most satisfying and the cheapest when looked at over the life cycle of the building, but also the most difficult to achieve. A typical building in this category would not necessarily be one that is constructed of the highest-quality materials but one that is designed carefully. Elevations would be relatively complex without being over-ornamented and, by blending the techniques of providing copings, cills, string courses and appropriate mixture of materials, the result would be a building that can 'carry' a certain level of staining without being unsightly.

5.4 SOILING, BUILDING COMPLEXITY AND AESTHETIC VALUE

5.4.1 Introduction

Based on aesthetic and perceptual studies of soiled and cleaned masonry buildings in Scotland, Andrews (1994) identified two types of soiling:

Figure 5.7 An older but associated building with similar stone cladding had already shown signs of uneven washing and deposition.

Figure 5.8 Detail of Figure 5.7 showing the sheltering effects of a protruding detail on the wall surface.

- *Consonant soiling* – this is where the soiling enhances architectural details of the building and, depending on the circumstances, possibly increases the aesthetic appeal of the building.
- *Dissonant soiling* – soiling that is unrelated to the building's architecture and is aesthetically displeasing.

According to Andrews, many plainer modern buildings cannot tolerate even the most minimal levels of staining and quickly become dissonantly soiled, whereas older buildings tend to accept similar amounts of soiling without much of a noticeable visual change. This concept has been developed to produce a more complex model of staining that tries to encompass a broader range of issues. It links three distinct concepts:

- aesthetic value
- soiling
- building complexity.

5.4.2 Defining terms

Soiling is reasonably self-explanatory and aesthetic value is similar to 'visual quality' of Verhoff's model. Building complexity is the interactional relationship between the characteristics of the façade and the amount of soiling present on it. For example, consider a traditional masonry building. As the façade is initially soiled it is consistent with the façade's architectural features and material texture. It becomes visually more pleasing as architectural features are emphasized (i.e. the façade becomes more complex) and its aesthetic value increases. If the soiling gets increasingly heavy, the underlying details are obscured, building 'complexity' is reduced, and aesthetic value drops until the whole face of the building is blackened and building complexity is at a minimum. The relationship between aesthetic value and soiling and aesthetic value and complexity for a stone

Figure 5.9 Same building after cleaning (see Figure 5.5 for comparison) with the former staining patterns clearly visible. It is predicted that algae staining will soon return.

Figure 5.10 Detail of Figure 5.9.

building is shown two-dimensionally in Figures 5.13 and 5.14. These have been combined to produce a three-dimensional model in Figure 5.15 which represents a hypothetical rubbled stone building. Only increased levels of soiling will force the profile into negative values.

This method of modelling seems to give a rational explanation of why many commentators see older, highly ornamented buildings as being more able to withstand the passage of time better than their modern equivalents. This is not because the ornamentation was designed specifically to throw water off the façade and prevent staining but that the nature of the façade materials and the complexity of that ornamentation meant that soiling could occur without negatively affecting the visual quality. Verhoff (1988) also acknowledges this effect and argues that in north European cities soiling is necessary to emphasize architectural detail because lack of strong daylight prevents the sharp, well-defined shadows that would otherwise do the job.

5.4.3 Application to new buildings

This three-dimensional model has been developed using older stone buildings as the subjects. The usefulness of applying this conceptual approach to new buildings might be marginal but it could give a useful insight into the importance of building complexity in a design of a façade. When considering some of the photographic studies in this book, a few assumptions may be made:

- most modern buildings have plain façades that lack architectural detail and ornamentation and so will never develop high levels of building complexity;

Figure 5.11 Cleaning the older building was slightly more successful but still did not return the façade to its original condition.

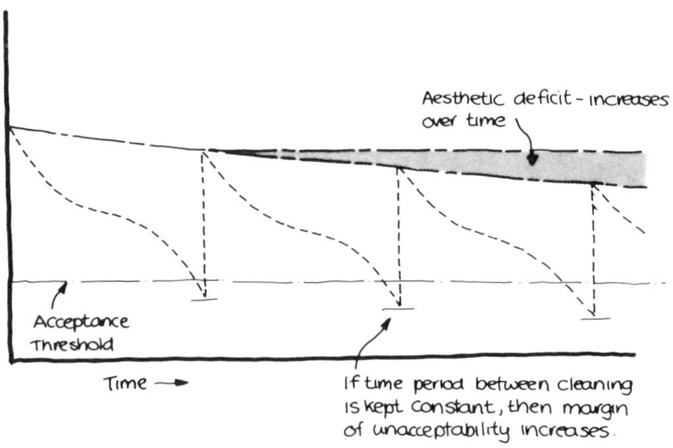

Figure 5.12 Amended graphical representation of design strategy 2 (see Figure 5.1) showing increasing aesthetic deficiency over time.

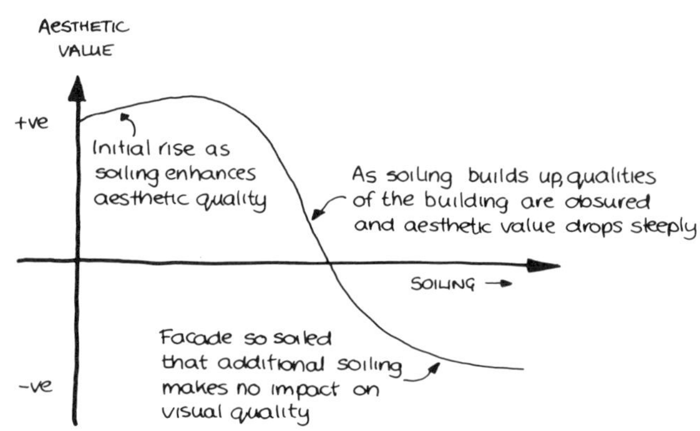

Figure 5.13 Relationship between aesthetic value and soiling (Andrews, 1994).

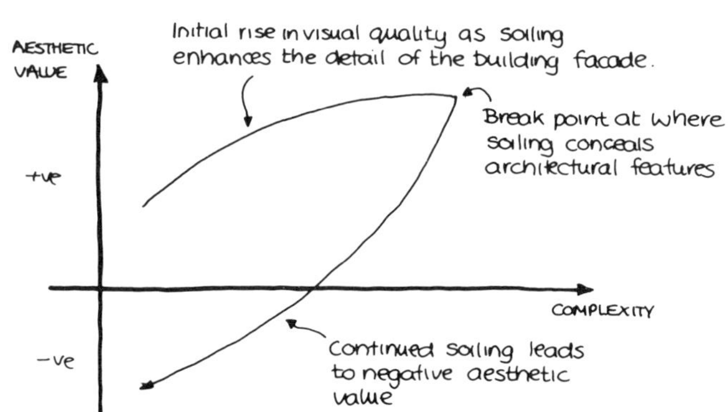

Figure 5.14 Relationship between aesthetic value and complexity (Andrewsm 1994).

- façades of non-absorbent materials will not be enhanced by light soiling;
- modern buildings are designed to achieve their full aesthetic value '*as new*' as opposed to traditional buildings where aesthetic values may increase in the first few years.

How useful these models may be in helping the designer with the very difficult task of producing a building that will perform well over time is questionable. But what they do offer are frameworks against which the designer can conceptualize the sort of end product that is to be produced. This puts the prevention of premature staining on the design agenda from the earliest stages, preventing it from becoming a last-minute detailing function carried out in a rush by the office technician just before the tenders are due to be sent out.

5.5 A SYSTEMS APPROACH

The third and final model, that of Atkinson (1977), attempts to make good the perceived deficiencies by defining the problem of staining in systems terms; identifying all the overlapping factors that will determine whether a building will stain or not. Four generic components were identified and are illustrated in Figure 5.16.

Figure 5.15 Relationship between aesthetic value, soiling and complexity (Andrews, 1994).

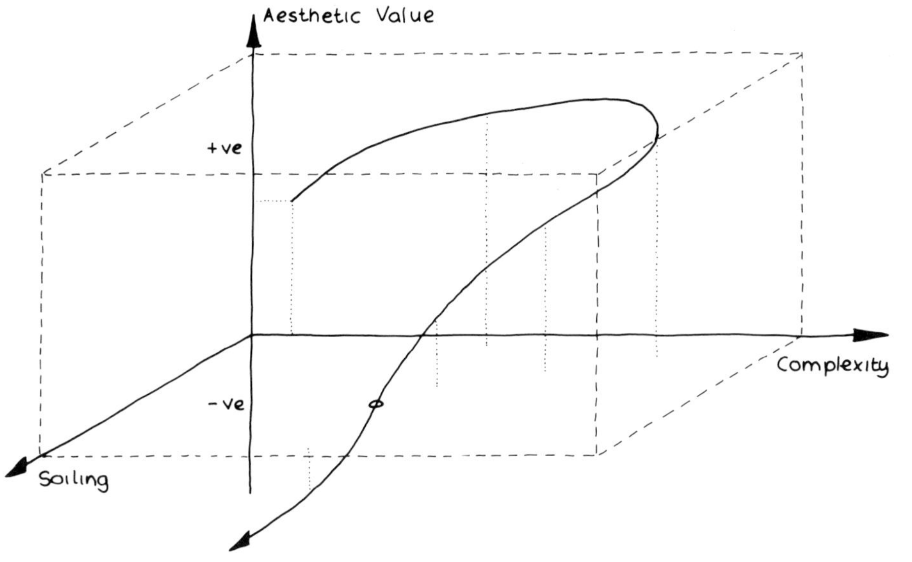

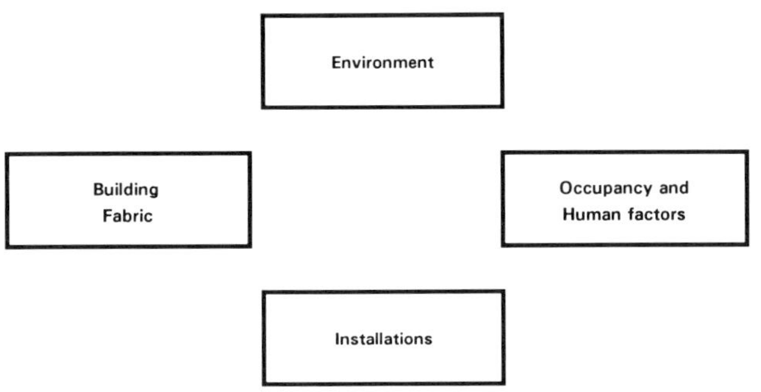

Figure 5.16 The four components of the performance and appearance 'system' (Atkinson, 1977).

Under each of these headings there are a whole range of other items and factors that have to be accounted for and only two of the components are under the designer's control – building fabric and installations. All of these issues are considered in more detail in the next section.

5.6 PREVENTION OF PREMATURE STAINING: A DESIGNER'S GUIDE

The model put forward by Atkinson is very thorough and the following section is an attempt to combine this framework with the knowledge gained from reviewing other publications and the observations of a large number of real buildings. It is hoped that the checklist produced will help designers identify the key issues that need to be resolved in order to minimize staining.

5.6.1 The environmental component

Climatic details. As part of normal feasibility studies for a building it is important to assess a site's climatic extremes as well as norms to get a feel for the full range of influences. Other factors to consider should include:

- *The direction of the prevailing wind* – full washing of elevations facing the prevailing wind is most common with a risk of staining/partial washing on more sheltered elevations.
- *North facing façades* – these are likely to remain damp for long periods and be vulnerable to the growth of algae and moulds. This is especially important with porous materials.
- *Rainfall levels* – this will help assess how damp the building is likely to remain.
- *Acidity of the rainfall* – this is important where the façade materials are affected by acidic solutions.

Local topography/ground surfaces. Does the local landscape expose or shelter any part of the new building above or below regional norms?

Height. As the height above sea level of the site will affect the distribution of rain on the façade, the dangers of partial washing of the plain elevations need to be assessed.

Building geometry. The shape and form of the building will affect the way air moves around it. The following features need to be reviewed:

- corners
- bays and setbacks
- podiums
- screens and low structures in front of the main building.

Consider how these might affect staining/washing patterns.

Environmental pollution. Assess the likely sources of pollutants in the locality and their likely effect on the new building. The following factors will influence this:

- the condition of the surrounding buildings will provide the most accurate insight;
- proximity to industrial plant and processes;
- proximity to main roads and busy rail networks;
- published information on pollution levels such as black smoke, SO_2 levels, air quality information, etc.

Because deposition can vary so greatly on a local level these indicators should be viewed very broadly.

5.6.2 The building fabric

The built form. The choices that have been made about the built form from the earliest stages of the design process need to be investigated. Factors to consider include:

- *Building shape* – will the shape of the building influence the way the elements act on it?
- *Height* – the height of the building can be as important as the elevation of the site in determining the potential danger of variable washing.
- *Surface configuration* – the shape and pattern of the surface can affect staining patterns. Inclined surfaces can be particularly susceptible.

Design details. The design of water shedding details is crucially important. Vulnerable features include:

- cornices
- string courses
- canopies
- plinths
- cills
- external stairs
- signs/lettering
- balconies
- hardstandings, etc.

The importance of careful detailing in avoiding building failures has been usefully discussed by Duell (1983) and can be applied to the prevention of premature staining. Duell argues that many designers give little priority to the process of detailing, preferring to leave it to inexperienced architects or technicians to sort out at the working drawing stage. Duell also points out that studies have shown that many architects are reluctant to refer to written technical information. To avoid these dangers a procedure for the development of constructional details was proposed and consists of six stages:

1. *Sketch details and outline specification*. In this stage the designer must bear in mind constructional details while developing the overall spatial and functional design by working back and forth between general and the detailed aspects without losing sight of the design.
2. *Selecting materials and components*. Because of the wide range of materials and components that are available, the designer may have to go back to first principles to analyse carefully the different components, because formal technical information is often biased, out-of-date or incomplete.
3. *Considering alternative materials*. By this stage some materials and components may have become closely associated with the design but it is still important to review critically their inclusion to see if it is a correct one. For those decisions that have not been made, a detailed investigation of the alternatives will help clarify and determine the choices. Costs can begin to influence component target costs and the client must be involved with decisions affecting functional performance, maintenance and the initial and life cycle costs.
4. *Development of sketch details*. Duell defines good detailing as 'the joining of materials and components in a functional and aesthetically pleasing manner'. Five common principles are offered that every detail shares in common:

   (a) *Function* – exclusion of elements, shedding of water, etc.
   (b) *Fit* – sizes and tolerances to avoid details which fail early.
   (c) *Practicality* – can it be done by reasonably competent operatives?
   (d) *Maintenance* – are details reasonable and practical considering situation and use?
   (e) *Costs* – are labour and material costs value for money?

   Duell makes an interesting observation that relates to staining problems: 'a detail should never be compromised for the sake of appearance. Recent architectural styles have resulted in details such as flush copings to parapets which have reduced the protection from the elements. The extent of the reported failures must lead designers back to "protective" detailing such as the provision of large overhangs at roof eaves'. Preliminary details should be sent to other design team members to ensure good coordination and complex details should be isometric to check that they are practicable.
5. *Final selection of products*. As the final details are refined, the specification for that feature should be gradually developing and intimately related to the design and detailing process. Materials and components may be specified for a number of reasons:

   (a) superior performance
   (b) preferred appearance
   (c) previous successful use, or
   (d) competitive price.
6. *Final critical appraisal*. It is important that details are critically appraised before they are issued as completed designs. This can be done by an experienced designer with a sound technical and constructional background and carefully structured to emphasize assistance and minimize criticism. Regular appraisals should be held to prevent abortive work.

Duell concludes that good detailing is one of the most important aspects in the design of buildings, helping to produce trouble-free buildings and satisfied clients. This applies equally to the problems associated with staining as well as other functional performance issues.

Surfaces and surface materials. The nature of the cladding material will determine whether they stain or not. Check the following:

- *Texture and colour* – will this highlight or mask dirt retention?

- *Chemical and physical properties and compatibility with adjacent materials* – will water running from one material affect another?
- *Presence of vegetation* – check the proximity of trees; the type of trees; planted areas; roof gardens; drainage of planted areas, etc.

5.6.3 Occupancy and other human factors

Design intentions and expectations. It is vital that clients' requirements are clearly identified at the briefing stage including:

- clients' views on costs and quality;
- image and status;
- the attitude of the local community;
- the input of planning authority, conservation bodies, etc.

Occupants and their activities. The following must be considered:

- local polluting activities associated with the building;
- accidental damage by processes carried out in and around the building, e.g. machinery, vehicles, industrial operations;
- vandalism and public attitudes towards the building – some building may be more of a target than others.

Building management. The commitment to cleaning and maintenance is vital. Issues to consider include:

- The organization of cleaning and maintenance – is it sufficient and appropriately funded?
- Adequacy of cleaning programmes – are they realistic and will they achieve the standards set?

If the façades require high levels of cleaning to retain their appearance, will the building owner have the continued commitment?

5.6.4 Installation

Rainwater collection. Are all areas that will receive high levels of run-off formally drained? The following need to be checked for adequacy:

- gutters
- downpipes
- gullies, etc.

Fixtures for cleaning and maintenance. Ease of maintenance and cleaning will influence how often both operations are carried out. Check the suitability of:

- ladders
- gantries
- cradles
- washing facilities, etc.

Check to ensure these fixtures are not causes of staining themselves.

Ability to service parts of building likely to cause staining. All the following installations could cause local staining problems:

- flues
- coolers
- water sprays
- air intakes and exhaust
- waste and soil pipes
- overflows
- oil supply lines
- disposal of solid waste.

In conclusion, Atkinson (1977) suggests that when designing a building the designer should account for all or at least most of these items listed above. For many projects this may be a daunting or impossible task as much of the required information may be difficult to retrieve or just not available. No matter how imperfect the process, Atkinson outlines three essential stages that a designer must go through:

- Know that there is an issue to be considered. In other words, premature staining of new buildings is a deficiency in the design process that needs to be corrected through reflection of past performance assisted by improvements in design training and practice.
- Obtain as much information as possible whatever the level of resources available.
- Know when there is no reliable information available.

In the latter case, Atkinson suggests that the 'experienced designer should play safe leaving the problem of advancement of understanding to the scientists.' The only problem with this approach is that, with respect to premature staining, the scientists do not seem to be advancing very fast!

REFERENCES

Andrews, A. (1994) Stone cleaning. A guide for practitioners. *Historic Scotland*, Edinburgh.

Atkinson, G.A. (1977) External vertical surfaces of buildings: aspects of design and appearance. *Evaluation of the Performance of Buildings*, RILEM, Espoo, Finland, pp. 19–25.

Duell, J. (1983) Developing a detail and specification. *Architects Journal*, 3 August, pp. 45–50.

Verhoff, L.G.W. (ed.) (1988) *Soiling and Cleaning of Building Façades*. Report of the Technical Committee 62 SCF, RILEM, Chapman & Hall, London.

Index